NEIGHBORHOODS

Volume 154, Sage Library of Social Research

RECENT VOLUMES IN
SAGE LIBRARY OF SOCIAL RESEARCH

112 Lidz/Walker **Heroin, Deviance and Morality**
113 Shupe/Bromley **The New Vigilantes**
114 Monahan **Predicting Violent Behavior**
115 Britan **Bureaucracy and Innovation**
116 Massarik/Kaback **Genetic Disease Control**
117 Levi **The Coming End of War**
118 Beardsley **Conflicting Ideologies in Political Economy**
119 LaRossa/LaRossa **Transition to Parenthood**
120 Alexandroff **The Logic of Diplomacy**
121 Tittle **Careers and Family**
122 Reardon **Persuasion**
123 Hindelang/Hirschi/Weis **Measuring Delinquency**
124 Skogan/Maxfield **Coping With Crime**
125 Weiner **Cultural Marxism and Political Sociology**
126 McPhail **Electronic Colonialism**
127 Holmes **The Policy Process in Communist States**
128 Froland/Pancoast/Chapman/Kimboko **Helping Networks and Human Services**
129 Pagelow **Woman-Battering**
130 Levine/Rubin/Wolohojian **The Politics of Retrenchment**
131 Saxe/Fine **Social Experiments**
132 Phillips/Votey **The Economics of Crime Control**
133 Zelnik/Kantner/Ford **Sex and Pregnancy in Adolescence**
134 Rist **Earning and Learning**
135 House **The Art of Public Policy Analysis**
136 Turk **Political Criminality**
137 Macarov **Worker Productivity**
138 Mizruchi **The American Corporate Network**
139 Druckman/Rozelle/Baxter **Nonverbal Communication**
140 Sommerville **The Rise and Fall of Childhood**
141 Quinney **Social Existence**
142 Toch/Grant **Reforming Human Services**
143 Scanzoni **Shaping Tomorrow's Family**
144 Babad/Birnbaum/Benne **The Social Self**
145 Rondinelli **Secondary Cities in Developing Countries**
146 Rothman/Teresa/Kay/Morningstar **Marketing Human Service Innovations**
147 Martin **Managing Without Managers**
148 Schwendinger/Schwendinger **Rape and Inequality**
149 Snow **Creating Media Culture**
150 Frey **Survey Research by Telephone**
151 Bolton **When Bonding Fails**
152 Estes/Newcomer/and Assoc. **Fiscal Austerity and Aging**
153 Leary **Understanding Social Anxiety**
154 Hallman **Neighborhoods**

NEIGHBORHOODS

Their Place in Urban Life

Howard W. HALLMAN

Volume 154
SAGE LIBRARY OF
SOCIAL RESEARCH

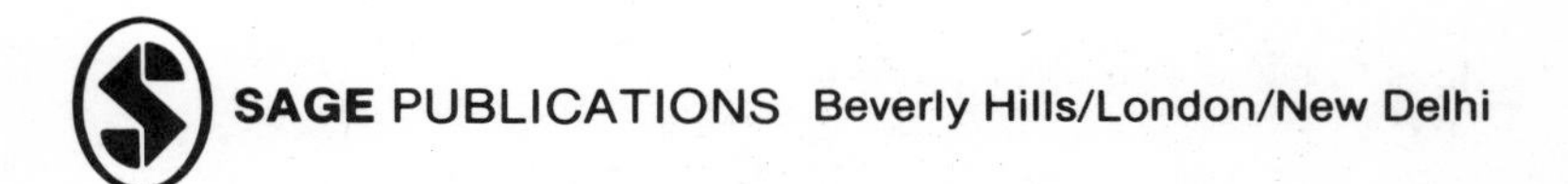

To Carlee

For information address:

SAGE Publications, Inc.
275 South Beverly Drive
Beverly Hills, California 90212

SAGE Publications India Pvt. Ltd.
C-236 Defence Colony
New Delhi 110 024, India

SAGE Publications Ltd
28 Banner Street
London EC1Y 8QE, England

Printed in the United States of America

Library of Congress Cataloging in Publication Data

Hallman, Howard W.
Neighborhoods : their place in urban life.

(Sage library of social research ; v. 154)
Includes bibliographical references and index.
1. Neighborhood--United States 2. Community organization--United States 3. Social action--United States. I. Title. II. Series.
HN90.C6H333 1983 307'.3362'0973 83-20024
ISBN 0-8039-2181-0
ISBN 0-8039-2182-9 (pbk.)

THIRD PRINTING, 1987

CONTENTS

PREFACE

Before I started to write this book, I had stated from time to time that my first neighborhood experience occurred one summer during my college years when I worked as a volunteer in a settlement house in New York's Lower East Side. There I saw and experienced the teeming street life and open air markets; shops on the ground floor, apartments above; schools, churches, social service agencies, and other institutions; everything the sociology textbooks described as neighborhood attributes.

But now, having been engrossed in ideas about neighborhoods for the year this book has taken to write, I have realized that my first neighborhood experience occurred as a young child in Pittsburg, Kansas, a city of 18,000. It was a small, personal neighborhood consisting of the houses within view and about one block in two directions where my playmates lived. When I entered elementary school my neighborhood became larger. It wasn't a bounded territory and didn't have a neighborhood association but rather was a section of town where I knew lots of children, knew of many adults, and was acquainted with the streets, alleys, vacant lots, and shortcuts. When I was nine, we moved one block to a different street, and my immediate personal neighborhood shifted. Play patterns for us kids tended to be linear along the street. So also for my parents social life. Though they had friends all around town, they regularly played canasta and staged an annual Fourth of July picnic with three other couples who lived within a block each way along the street. I guess I've always known intuitively that I grew up in a neighborhood, but I was later blocked by complex, academic definitions from acknowledging it rationally.

This isn't an autobiography, so I won't describe the other seven neighborhoods where I've lived and how my neighborhood experience has changed at different stages of life. Rather I want to suggest that if you, the reader, reflect upon your own life, your own experiences, and look openly at the experience of others, you'll discover that you already know quite a bit about neighborhoods and their place in your own life. You'll then be better equipped to study and understand various theories and different manifestations of neighborhood and to integrate the findings of research reports and descriptions of neighborhood programs. What you bring to reading the book is as important as all the studies I have synthesized in trying to present a comprehensive picture of neighborhood in the United States in the 1980s.

At different points in the book I emphasize that neighborhoods don't exist in isolation, that we as urban residents carry out our lives in several arenas of differing scale. For myself, during the year I wrote this book my office was at home in a Maryland suburb of Washington, D.C. (neighborhood-based employment!). Barbara Bush, who typed successive drafts on a word processor, runs an administrative services firm at a crossroads commercial area serving a community district of a number of subdivisions and small neighborhood clusters. I drew on the resources of 20 libraries located in my home county and in Washington and Baltimore, especially two city central libraries (Martin Luther King, Jr. in D.C. and Enoch Pratt in Baltimore) and the libraries of American University, the Urban Institute, and the U.S. Department of Housing and Urban Development. Using national network connections, I benefited from comments on the draft of the entire book by James V. Cunningham in Pittsburgh, Michael Bennett in Chicago, and Richard C. Rich in Blacksburg, Virginia. And also from comments on parts of the book by James Blackman, John Kromkowski, Arthur J. Naparstek, and Robert Zdenek, all based in Washington.

Beyond these reviewers, I am greatly indebted to numerous scholars who are partially recognized by references scattered throughout the text and listed in the bibliography. I also owe a lot to many practitioners with whom I've associated over the years and am now involved, though because the *do* rather than write, their contributions to my thinking can't be footnoted. They won't agree with everything I've written, but we share a conviction that neighborhoods are an important facet of American life and deserve loving, tender care.

— Howard W. Hallman
Potomac, Maryland

PART I

THE MANY FACES OF NEIGHBORHOOD

An Introductory Exercise

Before you read Chapter 1, write your own definition of neighborhood. This will give you an opportunity to compare your ideas with those of the author and others whom he quotes.

CHAPTER 1

INTRODUCTORY CONCEPTS

To be neighborly is a natural human trait. It combines our concern for others and our concern for self. As social beings, we relate to people around us. Though sometimes we are distrustful and unneighborly, we often help other people, and they help us. We sustain our lives through a division of labor and the exchange of goods and services. We give and receive, sell and buy, barter and trade. We gain security by enabling others to become secure. We love and are loved. A neighboring relationship is one manifestation of the interdependency of human life.

Likewise, the neighborhood is a natural phenomenon. It is organic, growing naturally wherever people live close to one another. It develops better in some environments than others, but can scarcely be eliminated completely as long as there are human settlements. Neighborhoods can be cultivated and nourished, protected and allowed to blossom to full maturity. Or they can be stunted, made to struggle for existence, but they will persist.

Neighborhoods are cells within larger settlements. That being so, neighborhood activities account for only part of urban life. How big a role the neighborhood plays in anyone's life varies among individuals. It also changes during different stages of life and even varies with the hour, day of the week, and the season. Neighborhood importance ranges from very little to very great. It never reaches zero, for even highly mobile people have some relationships with nearby people, services, and institutions. Nor does it ever amount to one hundred percent, for even the homebound depend upon food, clothing, and energy supplied from elsewhere. In other ways urban life requires interdependency far surpassing neighborhood boundaries, but still having important roles for neighborhoods.

PERSPECTIVES

If you and I look at neighborhood from our dwelling place outward, we note several things.

First, there is a hierarchy of scale. We have an immediate personal neighborhood in a small circle around us, demarcated by informal contacts with people living in nearby houses and apartments. Next comes a larger, functional area, perhaps centered on a school, church, or shopping area, or circumscribed by boundaries of organizations and service providers. Third, at least in larger metropolises, we have a loose identity with a community district, such as the Westside, Far North, or the Main Line suburbs, composed of a number of neighborhoods, and perhaps served by a community newspaper and a major shopping center. Beyond that there is the city or suburban county, and the whole metropolis. We live our lives in all these areas.

Second, we are linked to people and organizations within the neighborhood and beyond. We are part of networks of relationships. And we are served by functional systems whose interlocking operations range from small-scale, close-at-hand activities to large-scale enterprises covering a sizable territory. The role of neighborhoods can't be understood unless we also comprehend these linkages.

Third, we can conceptualize the neighborhood in different ways. This book uses five: as a personal arena, a social community, a physical place, a political community, and a little economy. Together they provide the whole picture. Part I considers these manifestations conceptually, and Part III approaches them practically, dealing with programs and activities carried out within neighborhoods. In preparation for the practical, Part II sketches a history of approaches to neighborhood action and delineates techniques of community organizing. Part IV draws together many strands by considering strategies to achieve neighborhood wholeness, but in a context where neighborhoods function in partnerships with wider spheres and are influenced and supported by outside forces.

Defining Neighborhoods

We should realize from the outset that neighborhood means different things to different people. There's no consensus among either residents or scholars on how to define neighborhood. That's because neighborhoods have many faces and take many forms. How they are perceived varies considerably. Into this uncertainty I plunge by offering a lean definition of neighborhood and then expanding it.

BASIC DEFINITION

A neighborhood is a territory, a small area within a larger settlement. It contains dwellings occupied by people, and usually community facilities and buildings with other uses. The residents come into contact with one another and interact. They are neighbors on the basis of propinquity, that is, nearness, regardless of any other social ties.

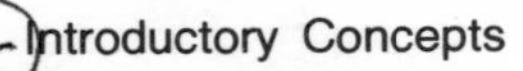

Therefore, a neighborhood is both a physical pla
community (minimally at least). In sum, *a neighbor*
territory within a larger urban area where people inha
interact socially.

By this minimum definition, all cities have neighbo
all dwellings in city and suburb are located in neighborhoods, and nearly everyone is a neighborhood resident even though he or she doesn't always articulate this role.

VARIED CHARACTERISTICS

Most neighborhoods have multiple features, though the degree to which a particular neighborhood possesses these varied characteristics ranges widely.

As a territory, a neighborhood is a *physical place*. If you say "I live in such-and-such neighborhood," your listeners visualize a physical setting. They can envision the houses, apartment building, schools, playgrounds, parks, churches, stores, maybe small factories, streets, and sidewalks. They have a sense of boundaries formed by expressways, railroads, rivers, and other natural features. But even if neighborhoods in different sections of the metropolis look alike to outsiders, to the residents their neighborhood has a distinct appearance and is differentiated from adjacent neighborhoods.

A neighborhood is a subjective entity as well as an objective reality. Its face and form and the social relations within are what individual residents perceive. An important determinant is the individual's informal neighboring activities and her or his patterns of going to nearby businesses and public facilities. Status factors and bonds of race, religion, ethnicity, or social class may affect the shape and size of the neighborhood a person identifies with. Each resident's *personal neighborhood of identity and use* varies, and it might be different for members of the same household.

As social beings, residents do things together. They share common interests and have similar values while still preserving opportunities for diversity and individual differences. They have a collective life carried out through social networks and sets of institutional arrangements. As this occurs, a neighborhood is *a social community*. Social ties may be even stronger if the neighborhood is populated mainly by a particular racial, ethnic, or socioeconomic group, though residents of heterogeneous neighborhoods can also achieve a strong sense of community. However, residents might also have conflicting values and clash over ous issues. Within a city and its suburbs, neighborhoods will show a onsiderable range in relative social cohesiveness.

A neighborhood contains housing to meet the fundamental human need for shelter. As a community, it is likely to have facilities to fulfill other functions of life: schools for education; yards, playgrounds, and recreation centers for leisure-time activities; churches and synagogues for religious observance; doctors' offices, clinics, and drug stores for health needs; grocery stores to supply food, and maybe some gardens; and other retail establishments. The list can go on and on, though the pattern differs from one neighborhood to the next. Thus, we have *functional neighborhoods* of varying degrees of completeness.

Many of these functions have economic characteristics, so a neighborhood can be perceived as *a little economy*. It has capital stock of buildings and community facilities. People rent, sell, and buy housing, so it's a miniature housing market. Commercial transactions occur at neighborhood shops and through saving and lending institutions. Many neighborhoods have companies and individuals producing goods and services, and this provides employment. Residents bring their earnings home and spend money within and outside the neighborhood, thus creating a neighborhood cash flow. However, neighborhoods vary in how well they articulate their economic life.

A neighborhood is also *a political community*. This can range from informal self-governance of a few aspects of neighborhood life to full-scale government, as is found in suburban municipalities of neighborhood size. The neighborhood is often a base of political action for dealing with governments of broader jurisdiction, such as city, county, state, and national. It might be a building block for voting, political parties, and representation in these wider domains.

Neighborhoods differ considerably in the extent they clearly express these basic characteristics: personal, social, physical, political, and economic. The most complete manifestation would be an incorporated enclave city or suburban municipality of neighborhood size with clearcut boundaries, its own civic center and community facilities, a good balance of employment and commercial services, a fairly homogeneous population or harmonious if heterogeneous, and local institutions controlled by the residents. But for each characteristic, a neighborhood might have a less complete expression: a neighborhood advocacy organization rather than self-government; not many local jobs and very few of the daily commercial services residents need; a scattering of community facilities so that some are located in adjacent neighborhoods; fuzzy boundaries; intergroup conflict; or apathy toward neighborhood problems. It's a matter of more-or-less, not either-or. Even though not complete, it is still a neighborhood according to our core definition: *a residential area of limited territory where social interaction occurs.*

EMPHASIS ON DIFFERENT FEATURES

Because of the many characteristics a neighborhood might possess, it isn't easy to divide a city into a complete set of nonoverlapping neighborhoods satisfactory to everybody. Thus, neighborhoods of personal relationships are usually small and are likely to have overlapping boundaries. Those of ethnic or racial identify have changing borders as population moves. To the extent an elementary school helps create a sense of neighborhood, shifts in enrollment, causing changes in school area boundaries, affect neighborhood interaction. A neighborhood commercial district might have a trading area not coinciding with other aspects of neighborhood life. Therefore, it is as hard to achieve agreement on specific neighborhood boundaries as it is to reach a consensus on how to define neighborhood in the abstract. But that doesn't disprove the existence of neighborhoods. Rather it forces us to live with a degree of indeterminancy. So be it.

The next five chapters prober deeper into the concepts of the major aspects of neighborhood. They elaborate on my expandable definition of neighborhood. But before moving on, let us look at how several others have defined neighborhoods. This may help you gain a better perspective of the many faces and forms of neighborhoods.

OTHER DEFINITIONS

In the mid-1960s Sociologist Suzanne Keller, while working for the Athens Center of Ekistics (headed by C. A. Doxiadis), surveyed the literature of sociology and city planning on neighborhoods. Although she found ambiguities associated with the term "neighborhood," she indicated (1968:87):

> Essentially, it refers to distinctive areas into which large spatial units may be subdivided, such as gold coasts and slums, central and outlying districts, residential and industrial areas, middle class and working class areas.

The distinctiveness can be attributed to four factors: (1) geographical boundaries; (2) ethnic or cultural characteristics of the inhabitants; (3) psychological unity among people who feel that they belong together; and (4) concentrated use of an area's facilities for shopping, leisure, and learning. However, the independent contribution of each factor is hard to assess. Keller noted that neighborhoods combining all four elements are very rare in modern cities, and in particular geographical and personal boundaries don't always coincide.

John McClaughry has incorporated similar elements into his definition, indicating that (1980:369):

> a neighborhood is defined as a predominantly residential area of a city that is (1) characterized by its own economic, cultural, and social institutions (schools, churches, police and fire stations, shopping districts, community centers, and fraternal and charitable organizations); (2) typified by some tradition of identity and continuity; and (3) inhabited by people who perceive themselves to be residents of the neighborhood and participants in its common life.

The National Commission on Neighborhoods struggled with definition and finally concluded (1979:7):

> In the last analysis, each neighborhood is what the inhabitants think it is. The only genuinely accurate delineation of neighborhood is done by the people who live there, work there, retire there, and take pride in themselves as well as their community.

Albert Hunter (1974) referred to such areas defined by residents as "symbolic communities." He described them as social objects with two main elements: cognition and sentiment. The cognitive element is displayed when residents give a name to their neighborhood and define, at least in their own minds, particular boundaries. Sentiment reflects their attachment to the neighborhood and their evaluation of its quality. (For how various individuals speak of their neighborhoods, see Watman, 1980.)

Roger S. Ahlbrandt, Jr., and James V. Cunningham have taken a functional approach (1979:9):

> The urban neighborhood, as it has unfolded, has come to take on many functions. These include the neighborhood as a community, as a market, as a service area, as a provider of shelter, as an arena for improving quality of life, as a political force, and as an actual or potential level of government.
>
> The neighborhood is a limited territory, containing an identifiable group of people and organizations. These people and organizations interact to some extent. They find certain of their needs taken care of within the limited territory. Sometimes some of them join together to take actions or improve the quality of life of the territory.

M. Leanne Lachman and Anthony Downs (1978) have emphasized shared participation and viewpoints among which are (a) use of the same space as a focal point for personal interaction, (b) a common relationship with nearby institutions, such as church or school, (c) a real estate entity, (d) a political entity, and (e) an instrument for exclusion. When Downs took up the subject several years later, he reviewed his earlier exposition plus definitions of the National Commission on Neighborhoods, Suzanne Keller, and others and boiled it down to a statement that

> *neighborhoods are geographic units within which certain social relationships exist,* although the intensity of these relationships and their importance in the lives of individuals vary tremendously (1981:15).

This is similar to my own core definition.

YOUR EXPERIENCES

Neighborhood is a personal matter as well as a topic for objective study. Therefore, I expect you the reader to bring in your own experiences and perceptions. I ask you to recall what neighborhood has meant to you at various times in your life, what it means now, and to project what it might mean in the future when you enter other phases of your life cycle. I hope also that you can imagine what neighborhood means to others of different ages, social classes, and geographic locations, and how they function in a neighborhood setting.

By combining the subjective and objective, you can gain a better understanding of the place of neighborhoods in urban life.

Exercises

(1) See how many definitions of neighborhood you can find.
(2) Identify the intellectual and political background and purpose (motivation) of the definers.
(3) Put these definitions on a matrix according to key traits. Notice which traits are emphasized and omitted by different definitions.
(4) Now again write you own definition, as you did before reading Chapter 1.
(5) Before reading Chapter 2, identify the boundaries of your present neighborhood. List the ways this neighborhood has meaning to you and describe how you use it. Do the same for other neighborhoods where you have lived in earlier periods of your life, going back to childhood. Talk to your friends about how they perceive their neighborhoods, present and past.

CHAPTER 2

A PERSONAL ARENA

As an urban resident, each of us has a personal neighborhood. There we have neighborly relationships with nearby residents, and there we are able to fulfill some of our human needs. Acquaintance with neighbors may range from scant to full, and our reliance upon the neighborhood for goods and services may vary from little to great, but it is a rare urban dweller who has no ties at all with people, institutions, and businesses in the vicinity of home. The bounds of our personal neighborhood may differ from those of others, even from others within our own household, but nevertheless most of us feel an identity with a small territory divided out from the larger city and metropolis, that is, with our neighborhood.

Yet, the neighborhood isn't ours alone, for it belongs to other people, too. It has institutions serving us collectively. We interact with one another, cooperatively and competitively, form social networks, and together create a community with its own existence and continuity. Individuals move in and out but the community persists, though it changes over time.

In this chapter we examine the personal neighborhood from the individual's perspective — cognitive and functional. In the following chapter we deal with the collective and analyze neighborhood as a social community.

OUR CYCLE OF LIFE

Many variables affect individuals' perception of their personal neighborhood: age, family status, socioeconomic class, race or ethnicity, social values, life styles. Physical traits of the neighborhood are also factors, such as housing types, building density, physical condition, location within the metropolis. Of these variables, the individual's age is one of the most significant.

A child of three or four, just beginning to gain freedom from the confines of home and own yard, has a quite small neighborhood: a few houses in either direction along the street, or one play area of an apartment project. Her or his playmates may have contacts with other children and adults in different constellations and therefore somewhat different personal neighborhoods. As children grow older, their neighborhoods expand: across the street, around the corner, into the next block. When they enter school, mobility increases, friendships widen, and their neighborhood grows proportionately.

During the latter years of childhood, say from eight to eleven or twelve, the neighborhood takes on many rich associations. Indeed, many of us have vivid memories of our neighborhood of those years, for we spent most of our time there: within the immediate block or two of our residence and within the larger territory of the elementary school district. We played there, attended school, went on errands for our parents, knew the vacant lots, alleyways, the shortcuts. We were acquainted with many adults and knew others by reputation. Some of us earned our first money doing odd jobs within the neighborhood. It was our world, or at least a large hunk of it (see Berg and Medrich, 1980).

Adolescence opens new horizons. Because secondary schools are larger, they draw on several neighborhoods. Even beyond school, teenagers have greater mobility. They branch out and explore other parts of the city. At first the jobs they hold may be near home, but after a while many go beyond the neighborhood for employment opportunities. When they reach the legal driving age, their orbit expands even wider. Those too poor or too city-bound to have a car in the family usually have access to public transportation to go afield. Nonetheless, neighborhood remains part of most teenagers lives. It may be the site of much of their leisure-time activities — organized events and just hanging around. Peer and subcultural influences are neighborhood-based. Sometimes youths form gangs within their neighborhoods and become very turf conscious.

Older youth and unmarried young adults have the least amount of neighborhood consciousness of any segment of society, unless they are anchored in a racial or ethnic community encompassing a large part of their lives. Those in college may have moved away from their home neighborhood, though the dormitory or student apartment building may function as a temporary neighborhood. Those in the armed services or in prison probably find the barracks, ship, or cellblock not much of a substitute neighborhood. The ones employed usually work outside their immediate neighborhood, and they roam around for social activities. Most young adults are unlikely to become very much involved in neighborhood civic and political activities. Nevertheless, they, too, have an immediate neighborhood of use: grocery store, dry cleaners, laundromat, perhaps church or singles bar. And to some extent they have neighborly relationships with other residents in their building.

In the contemporary United States the vast majority of young adults live in a house or apartment separate from their parents by the time they are 25. As they settle down and rent or buy their own dwelling unit, they display a greater awareness of neighborhood location: its nearness to kin, their parents' home, or social group; convenience to work; facilities and institutions supporting their life style; for those wishing to rear children, the school situation; social norms and values held by residents; what they can afford balanced against the social status of the neighborhood. Not all of these factors influence every young adult seeking a house or apartment, but something about the neighborhood is likely to influence their choice.

Adults with children are among the most neighborhood-conscious Americans, both in deciding whether it's a suitable neighborhood (regardless of how they define "suitable") and in tying into neighborhood networks (Fischer, 1977:146). Even so, among adults their use and neighboring patterns vary considerably. Those remaining in the neighborhood during the day often carry out a considerable number of life's activities there, especially those with young children. Single parents or families with both parents working may seek day care or after-school supervision for their children within the neighborhood. Neighboring, shopping, leisure time pursuits, congregate religion, civic endeavors, politics, and medical care are among the activities many adults fulfill at least in part near home, though with varying intensity and completeness.

Older persons, as physical strength and economic resources ebb, become more dependent on people, services, and institutions near home. They may move to smaller quarters, perhaps in a different neighborhood. As aging proceeds, their immediate neighborhood shrinks, opposite to the expansion experienced by growing children (see Millas, 1980).

Thus, for each of us as individuals the face and form of our personal neighborhood changes throughout our life. (For a British version, see Mann, 1970).

TIME/SPACE

In addition, shorter variations in time also affect our personal use of the neighborhood.

School-age children traverse the streets in the morning and are all over the neighborhood after school and on weekends. Nonworking mothers may be about the neighborhood more during the day when their children are in school. Retired persons are likely to be away from their homes more during the day than at night. Residents employed outside the neighborhood use its streets and sidewalks going to and from work, gather at bus stops in the morning, stop at neighborhood shops on the way home, and are around more on weekends. By way of illustration, Figure 2.1 presents a daily space and time prism for a two-adult household.

There are also annual variations. During the nice weather months, more resident are in their yards and on the street. In the hot summer months in low-income areas with crowded apartments lacking air conditioning, residents may remain outside very late at night. But in more affluent neighborhoods in summertime, people stay inside air-cooled homes, and many go away for the weekend and for vacation periods, making it harder to hold neighborhood meetings. It's also more difficult to get a good turnout on very cold, wintry nights. Indeed, time of day, day of the week, month of the year affect how we use our personal neighborhood.

PERSONAL RELATIONSHIPS

As social beings, each of us has personal relationships with other people. Depending upon our mobility and our personality (how outgoing or ingrown we are), the number we relate to personally might range from a few to hundreds. They might include relatives (within our household and living elsewhere), neighbors, friends, coworkers, fellow students, and persons with whom we share particular interests. Also, we meet people because of their functional role — shopkeepers, clerks, customers, teachers, clergy, politicians, recreation leaders, and many more — and then we develop a more personal relationship with some of them.

"Dyadic ties" is the term anthropologists use to describe these two-person relationships (Eames and Goode, 1977:119). They form our initial link with social networks. Some are neighborhood-based, some are not. Of particular interest to us in this chapter are relationships which individuals have with relatives, friends, neighbors, and interest-sharers within their neighborhood.

Suzanne Keller has made a useful distinction between relative, friend, and neighbor (1968:24-26). You gain relatives by birth and marriage. Friends you choose. Neighbors are present in nearby space. Friends and relatives may live anywhere, but neighbors by definition always reside near at hand. You are acquainted with your neighbors, usually know their faces and names, and may offer mutual aid in certain circumstances, but merely as neighbors you don't have the affection and

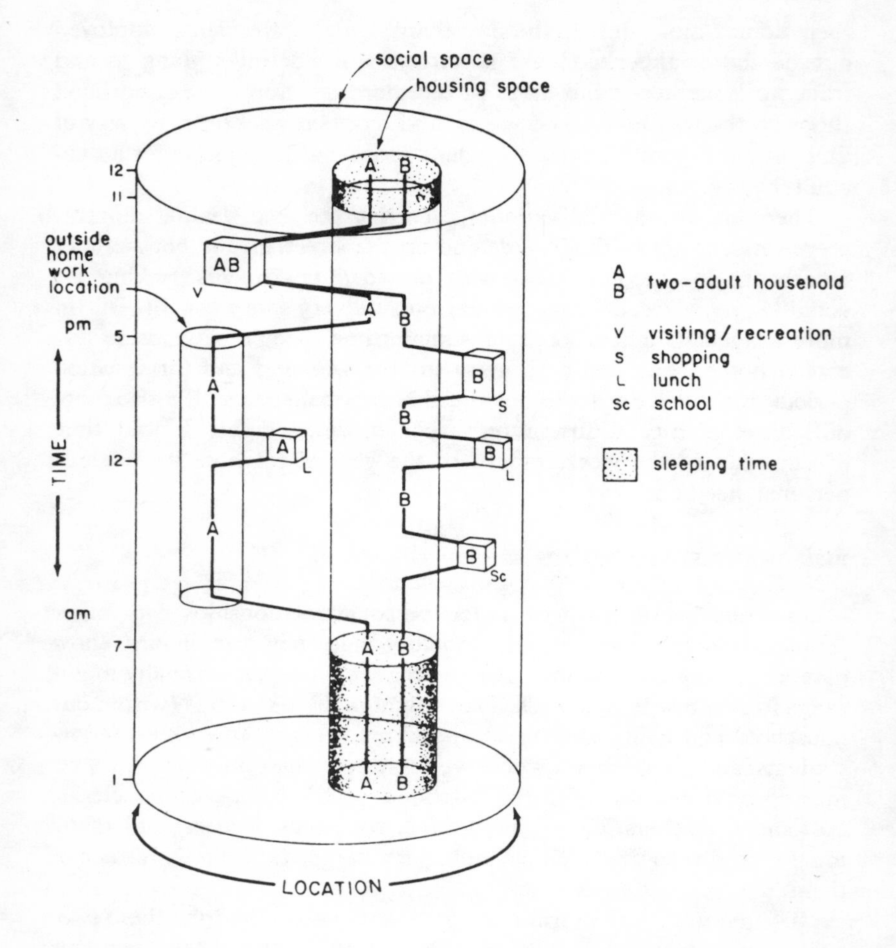

From *Internal Structure of the City: Readings on Urban Form, Growth, and Policy*, second edition, edited by Larry S. Bourne. Copyright © 1971, 1982 by Oxford University Press, Inc. Reprinted by permission.

Figure 2.1 Daily Time-Space Prism of Activities of a Two-Adult Household

close bonds which characterize friendship. You might become friends with your neighbors (thus adding depth to the relationship) and friends and relatives may live nearby (thus adding another dimension to those relationships), but proximity alone is the dominant factor in forming the neighbor relationship. Friendships continue when people move and you are seldom completely out of touch with relatives (at least to the level of grandparents, uncles, aunts, nieces, and nephews), but neighbors lose contact immediately when they relocate (unless they have become friends).

To this threesome I add a fourth: persons with whom we share common interests. Our focus might be music, arts, recreational activities, religion, vocation, a cause. We may have come together originally because of this interest, or we may be acquainted for other reasons and find we have a strong, mutual interest. This kind of sorting out occurs within a neighborhood as residents come in contact. Persons sharing a common interest get together for that purpose. In contrast, neighboring is initiated through propinquity, and friendship is built upon a broader set of mutual values and bonds of affection. Neighborly contacts may lead to interest sharing, and interest-sharers may become friends, but by itself interest-sharing concentrates upon the single dimension of the common concern. The neighborhood itself may be such an interest and can bring together people who weren't previously acquainted either as neighbors or friends.

Also, as Ulf Hannerz has noted (1980:264): "someone who is durably present in a neighborhood because of his work is in a sense also in a neighbor role, as not every encounter between him and the others there need involve a work task on his part." This includes shopkeepers, delivery personnel, social service workers, clergy, and other professionals. Thus, neighboring can occur among residence-neighbors and work-neighbors.

NEIGHBORING

Activities people carry out in their roles as neighbors can be called *neighboring*. According to Keller (1968:44), neighboring has the following dimensions:

(1) Neighboring is a socially defined relationship ranging from highly formalized and institutionalized rules and obligations to highly variable, voluntary exchanges.

(2) In essence, neighboring involves exchanges of services, information, and personal approval among those living near one another, however nearness is defined.

(3) The needs prompting these exchanges may be divided into four categories:

   (a) The daily, unexpected occurrence that is unforeseen yet recurrent, such as running out of bread or needing to post a letter.

   (b) The big emergency, such as a fire, illness, or death.

(c) The significant collective event — marriage, birth, a holiday.

(d) Cyclical collective needs — as at harvest time, during economic depressions, and during job layoffs. The aid exchanged among neighbors is both material and spiritual.

(4) Neighboring exhibits varying degrees of intensity and frequency although our knowledge here is extremely meager.

She also pointed out that neighboring has both manifest and latent functions. Manifest functions "involve the exchange of moral and material aid, including tools, information, and advice, in times of minor and major crises." Latent functions include (1) the exercise of reciprocal social control to help sustain common standards and shared communication, (2) the supply and spread of information, quickly and efficiently, via gossip, throughout a given area, and (3) the creation and maintenance of social standards or correct belief and conduct. "Only by communicating with others," Keller observed (1968:45), "do individuals learn the accepted rights and wrongs of their milieus."

The extent any person is involved in neighboring ranges widely, influenced both by individual preferences and the social milieu in which the person resides. There are also differences to the degree that neighbor acquaintances become friends. Some housing designs bring people into greater contact than others, such as courtyards and recreation areas of apartment projects, proximity to sidewalks and streets of rowhouses, open yards of single family dwellings, shared driveways and alleys (more on this in Chapter 4). Yet, the crowdedness of high density apartments might drive occupants inward to obtain privacy; indeed, there is probably less neighboring there than in any other setting (McGahan, 1972; Zito, 1974).

Personalities range from shy to gregarious, and this affects how neighborly one becomes. Time spent within the neighborhood is a factor, such as the contrasts of housespouse and working parent, child and older youth, seldom-away elderly and seldom-home young adult. Persons absorbed with relatives and close friends or with their work and special causes have less time for neighbors. Neighborhood ethos itself is a major influence because of the demands and expectations it creates, such as the circumstances and conditions of mutual aid and whether residents are expected to be outgoing or restrained in their neighboring. People with similarities tend to achieve closer neighbor relationships, and this becomes particularly noticeable where the neighborhood is also an ethnic or minority racial community. However, heterogeneity doesn't preclude neighborliness.

In her 1968 book Keller, having defined neighboring in terms of material and moral support, found evidence in studies around the world for a decline in neighboring as people moved from rural to urban settlements and as metropolitan society grew in size and complexity. She offered the following reasons for this decline (1968:58):

(1) The presence of multiple sources of information and opinion via mass media, travel, voluntary organizations, and employment away from the local area.

(2) Better transport increasing mobility beyond local village or district boundaries.

(3) More differentiated interests and desires as well as differentiated rhythms of work resulting in a lowered inclination to neighbor selectively. This also results in lesser amounts of shared free time available for leisure.

(4) Better social services and greater prosperity and economic security.

"These changes," she concluded, "affecting solidary working class districts in cities no less than solidary small towns and rural villages, result everywhere in a decline of organized as well as spontaneous neighborly associations."

This, however, is a premature conclusion about both the extent of neighboring and, as we shall see later, neighborhood associations. Research data show otherwise. For example, surveys in city and suburban neighborhoods in the Detroit area in 1974-1975 revealed that 56 percent of persons interviewed said they had been helped by neighbors in some kind of life crisis situation. Although one-fourth of the respondents said they had no friends in the neighborhood, 36 percent of this group called upon neighbors for help (D. Warren, 1981:60). A 1980 telephone survey of a cross-section of Pittsburgh residents found that 51 percent visit often or sometimes with neighbors, 60 percent help or are helped by neighbors with small tasks, and 95 percent feel they can call on neighbors for help in an emergency (Ahlbrandt and Cunningham, 1980:60). Other studies affirm the persistence of neighboring among various racial, ethnic and socioeconomic groups in many different locals (Tomeh, 1967; Nohara, 1968; Suttles, 1968; Litwak and Szelenyi, 1969; R.A. Wilson, 1971; Williams, Babchuk, and Johnson, 1973; Hunter, 1974; Antunes and Gaitz, 1975; Martineau, 1977).

For most of us, informal neighboring takes place in a fairly limited territory. Maybe with the occupants of the houses on either side, the one in back, and a couple of houses across the street. Or the people along our walkway or stairwell of garden apartments. The ones from our tenement with whom we sit on the front stoop, or those we meet at the apartment building's swimming pool. We exchange pleasantries, information, and on occasion help each other on some matter. Our neighbors in turn have their own circle of neighbors, extending outward to people we scarcely know. Circles overlap and through a network of social contacts form a larger neighborhood. But for us as individuals our locale for neighboring encompasses a very small area. The face-block, containing dwellings on the opposite sides of a street, is an approximation of this immediate, personal neighborhood.

FULFILLING FUNCTIONS

But there is another aspect of our personal neighborhood which is larger geographically: the one in which we fulfill some of life's basic needs close to home. Shelter, of course, is a neighborhood function by the very definition of neighborhood. Beyond housing, many other functions of life are carried out in neighborhoods, although individual differences are apparent.

These days most births in the United States occur in hospitals outside the neighborhood, but child rearing takes place in homes and within a neighborhood setting. Parents have the basic responsibility, perhaps aided by other adults in an extended family and assisted along the way by babysitters and sometimes day care personnel. Often there is informal support from neighbors in the arduous task of child rearing. As babies develop into children, they become playmates with neighbor children under the protective eyes of parents and other neighborhood adults. Their territory expands as they get older and gain greater freedom of movement. Yet, child care during the first years of our children's lives occurs almost exclusively within the orbit of an immediate, personal neighborhood.

Learning starts the day of birth. Formalized it is called education and takes place in schools. For some this commences with nursery school, many more with kindergarten, and for virtually every child at least by first grade of elementary school. Mostly these schools are nearby, but exceptions occur when parents choose private schools, public policy demands busing to schools outside the neighborhood to achieve racial integration, and the necessity of drawing together a school-size population requires busing in low-density suburbs and rural areas. Even with these exceptions, most children attend elementary school in or near their own neighborhood, and their parents are involved with teachers and other school personnel in fostering the educational process.

Recreation is a neighborhood function, especially children's informal play and both free-form and organized activities at public playgrounds and schoolyards. Although teenagers roam outside the neighborhood, many spend a great deal of their leisure time in or near their homes. Organized youth activities often have a neighborhood base. So, too, do adult activities, whether organized by friends, church, club, or agency. Although the neighborhood doesn't have a monopoly on recreation, it is a significant site for diverse leisure-time activities.

Most meals are eaten at home, but most food comes from outside the neighborhood. Grocery stores are the connecting link. At one time they were mostly neighborhood-based, but no longer, especially in medium to low density suburbs. Nevertheless for many, shopping for food, household items, and some clothing, and also obtaining such services as dry cleaning and appliance repair, occur at least partially in or near the neighborhood.

For many, congregate religious expression takes place in the neighborhood, especially for Roman Catholics and Episcopalians organized into geographic parishes and Orthodox Jews, who are required to walk to their synagogue. Other Jews, Protestants, and believers of other faiths are less confined to their neighborhood, though most probably go to a nearby synagogue or church of their denomination.

Urban living requires a host of public services, many of them delivered within neighborhoods, and often, though not always, organized on a neighborhood basis (as is reviewed in Chapter 12). They include such services as police, fire, refuse collection, street maintenance, public recreation, libraries, housing inspection. Residents call upon these public personnel to obtain particular services to satisfy specific needs.

But as urban dwellers we don't rely exclusively upon government, for we help one another through exchange of skills, emergency assistance, and mutual protection. As caring neighbors, we look out for the elderly, children, youth, and troubled adults. We organize neighborhood associations, mutual benefit societies, clubs of many varieties, cooperatives, protest committees, and political action units. In this manner we work together to carry out activities we consider necessary.

As individuals, we have our own patterns for fulfilling these functions of life near home. The territory in which this occurs is unique for each of us. This varies not only according to our preferences and life styles but also to the extent that services and facilities we need are located near by. The more complete a neighborhood district is in available services the more likely will the personal functional neighborhoods of the residents coincide.

In a citywide survey conducted in Chicago in 1967-1968, Albert Hunter (1974: 130, 133) asked people where they go for certain activities. He found that 73 percent went to church in their neighborhood and 71 percent did most of their food shopping there, but only 20 percent went mostly to neighborhood places for recreation and only 15 percent worked there. Hunter also learned that 37 percent belonged to at least one organization within the neighborhood and that more than half of the non-joiners were aware of voluntary associations active in the neighborhood.

Ten years later Ahlbrandt and Cunningham (1980:80) checked out patterns in Pittsburgh and discovered that frequent use of facilities in or near the neighborhood was as follows: grocery shopping, 63 percent; attending church or synagogue, 51 percent; using health or medical facilities, 45 percent; use of recreation facilities, 20 percent. Of the respondents 56 percent were aware of an organization in their neighborhood dealing with neighborhood problems, and 17 percent belonged to at least one such organization; and 35 percent were members of a church, PTA, or fraternal organization in or near their neighborhood.

The use of neighborhood facilities may differ in other cities and regions, and even among neighborhoods within the same city, but it occurs everywhere. Virtually all urban inhabitants live at least part of their lives within the orbit of their own personal neighborhood: a quite limited territory where neighboring occurs and a larger area, still near home, where they can fulfill some of life's basic functions.

SECURITY AND STATUS

Our personal neighborhood serves other purposes in our lives. In particular it can offer security and provide status.

Urban life has many hardships, especially for the poor and those on the edge of economic insecurity. But even the more affluent experience the competitive pressures of work and the marketplace, the stress of congestion and rushing about, and feelings of being isolated in a crowd. All segments suffer from environmental deficiencies, and all face threats to personal safety.

In this setting, home — extended to encompass the neighborhood — is a safe harbor. It is a place where you want to feel at ease, put aside some of your worries, relax, be comfortable with people you know and trust. This is a major reason that most neighborhoods house people with similar characteristics and contain a fairly limited span of socioeconomic groups and life styles. To Andrew M. Greeley, this tendency accounts for ethnic concentration in particular neighborhoods (1977:88):

> They choose their homes in areas relatively close to where they lived before and where families and friends tend to live; they also choose neighborhoods by the availability of community services which they have grown up to believe are important (synagogues, parochial schools, art centers, and so forth). Without reflecting much on the ethnic composition of a neighborhood, they end up more likely in neighborhoods with their own kind. If you ask them whether they had been "looking" for an Italian, Polish, or Irish neighborhood, they may deny it; what they were looking for was a neighborhood in which they would feel comfortable. For a number of conscious and unconscious reasons, it often happens to be a neighborhood where a substantial number of their own kind live.

The same kind of preference for a comfortable neighborhood comes out of a set of surveys in the metropolitan areas of Boston, Rochester, Dayton, Houston, and Kansas City, Missouri, on people's attitudes toward neighborhood and how they choose where to live (Coleman, 1978). Significant determinants included social comfort of adults, influence on children, and physical safety. Interviewers found that a substantial majority agreed with the statement "You have to approve of the neighborhood before you ever consider a house." The higher the person's socioeconomic status, the more strongly was this position taken. Likewise in another survey in Greensboro and Durham, North Carolina, the vast majority said they would consider the neighborhood first, the house second (R. L. Wilson, 1962).

Social status is an important factor in determining neighborhood suitability. Most people want to live in the highest status neighborhood they can afford, though one not too different from their own life-style and with people not too different so as to be uncomfortable. Neighborhood identity is shorthand for defining one's status to others, for metropolitan residents carry a rough status map in their heads (Ross, 1962). Your neighborhood becomes a symbol for your social ranking.

While this holds true with the majority, there are exceptions of families and individuals who ignore status and seek greater heterogeneity than usual or move to housing in less safe, less prestigious neighborhoods than they can afford. They include the affluent, return-to-the-city group, especially those without children, who have refurbished rundown houses near downtown. Gentrification is the term sometimes used for this process, or displacement when seen from the perspective of lower income occupants who can't afford the rising prices. But even this pioneering gains the newcomers some subtle status among friends in the know.

NEIGHBORHOOD- OR CITY-ORIENTED

Yet, how important is one's personal neighborhood compared to the broader city with all its opportunities? Social scientists have offered polar positions on this question.

One tradition traces back to Ferdinand Toennies' *Gemeinschaft und Gesellschaft* (1957), a pair of ideal types translated as community and society, or communal and noncommunal. In our context, the former is manifested in the neighborhood and the latter in the metropolis, or as "urbanism." Louis Wirth (1938:20-21) used the latter term to describe the urban mode of life "consisting of the substitution of secondary for primary contacts, the weakening of bonds of kinship, and the declining social significance of the family, the disappearance of the neighborhood, and the undermining of the traditional basis of social solidarity." Subsequently sociologists have studied what portion of the population could be considered as "urbanites" and have debated Wirth's thesis (Foley, 1952; Greer, 1956, 1962; Bell and Boat, 1957; Wilensky and Lebeaux, 1958; Gans, 1962b; Guterman, 1969; Kasarda and Janowitz, 1974; Hunter, 1975).

Another tradition goes back to Charles Horton Cooley (1909), who stressed three primary groups — the family, the play-group of children, and the neighborhood or community group of elders, "characterized by intimate face-to-face association and cooperation" (1909:23). In our era Peter L. Berger and Richard John Neuhaus have written of four mediating structures — neighborhood, family, church, and voluntary association, "defined as those institutions standing between the individual in his private life and the large institutions of public life" (1977:2). Andrew M. Greeley has argued this case by insisting that

> there is a strong disposition in most humans to be gregarious, and "industrialization" and "modernization," far from eliminating "neighboring," as some theorists of the mass society would have us believe, provide more time, more occasion and more resources for it, and a much more sophisticated vocabulary for talking about it. It is, in other words, virtually impossible to prevent most people from setting up close interaction networks among those who live close to them (1977:85-86).

The most recent empirical evidence comes from the Ahlbrandt-Cunningham survey in Pittsburgh (1980). In conducting telephone interviews with a sample of 5,896 persons from all sections of the city, they discovered that nearly two-thirds (63 percent) felt more loyalty to their neighborhood than to the city. But there was considerable variation among neighborhoods, ranging from 39 to 92 percent of the residents who held stronger loyalty to their neighborhood. Other parts of the survey, previously quoted, indicated different patterns in use of neighborhood facilities for grocery shopping, workshop, health care, and recreation. Of the respondents 72 percent felt that their neighborhood was a good or excellent place to live, but relatively few of them relied upon it exclusively for all of life's functions. Similarly in Chicago, of Hunter's respondents who expressed their feelings about their neighborhood, 65 percent were positive but they too went about the city for various activities (1974:118).

Unfortunately, as Joseph R. Gusfield has observed (1975:13), "In current sociological usage, the concepts of 'community and society' have emerged as opposites in an almost zero-sum form. That is, whatever accentuates society diminishes community, and vice versa." Yet to pose neighborhood versus city orientation as an either/or choice makes little sense in real life. Each of us is more-or-less neighborhood oriented for certain parts of our lives and more-or-less city, regional, national oriented for other concerns. We differ in this combination during various stages of our life cycle, time of day and week, and the season. Moreover, there are variations related to marital status, occupation, and socioeconomic class. The pattern will vary among cities, regions, and between city and suburb. Nevertheless, for most of us Americans, our personal neighborhood holds an important, though not exclusive, place in our lives.

Exercises

(1) Make a time-space chart of your activities during a typical week. Which activities occur within your neighborhood? Which ones occur elsewhere?

(2) Make a similar time-space chart for earlier periods in your life when you lived in a different neighborhood or used your neighborhood differently.

(3) Chart your major personal relationships. Categorize each person as kin, friend, neighbor, fellow student, coworker, interest-sharer, service provider, customer, or other. How many of these one-to-one relationships occur within your neighborhood? Where else do they occur?

(4) List the acts of neighboring you have been involved in during the past month.

(5) **Try to find someone (or imagine someone) who has no neighborhood connections and determine to whom he or she relates. Do you think such persons are rare or numerous?**

CHAPTER 3

A SOCIAL COMMUNITY

The personal neighborhood is rooted in relationships of people with one another, in nearby facilities they use, and in the local organizations in which they participate. When we look at this phenomenon from a broader perspective, we observe that personal neighborhoods overlap, interact, and combine. People get together in common endeavors, and they relate to specific institutions and agencies organized on a territorial basis. A social fabric bonds the people to one another, and shared values and norms influence their lives. In short, the collective entity called a neighborhood is a social community.

WHAT IS A COMMUNITY?

To call a neighborhood a community introduces another concept as elusive in definition as neighborhood. That's because community means different things in different contexts. Thus, George A. Hillery, Jr. (1955) found 94 definitions of community, though Joseph R. Gusfield (1975) has noted that the usage falls into two main categories: territorial and relational.

In attempting to gain an understanding of the metropolitan scene, a social science panel of the National Research Council focused on community as "a territorially bounded social group" and offered this definition (1975:2):

> A community consists of a population carrying on a collective life through a set of institutional arrangements. Common interests and norms of conduct are implied in this definition.

They realized that the concept has both micro and macro meanings. As to micro:

> One use of the word community then is to refer to a grouping of people who live close to one another and are united by common interests and mutual aid. In this sense, a community is small numerically, consisting of, at most, a few hundred people, and the connotation is one of solidarity.

In contrast is the macro community:

> On the other hand, the term may be used in the broader sense to refer to any population that carries on its daily life through a common set of institutions. In this sense, it may apply to a population aggregate of any size, for example, one in which the members participate in the division of labor within a particular socioeconomic system. The emphasis, in this instance, is on the interdependence that stems from specialization and exchange.

Roland L. Warren (1978:9) has linked the functional with the territorial, indicating that a community is

> *that combination of social units and systems that perform the major social functions having locality relevance.* In other words, by *community* we mean the organization of social activities to afford people daily local access to those broad areas of activity that are necessary in day-to-day living.

Yet Warren and the National Research Council panel have realized that community has other meanings. The latter noted (1975:3) that

> community is used increasingly to refer to interest groups whose common activities are relatively independent of location factors. The source of the common interest may be artistic or scientific, commercial or governmental, religious or ethnic.

This can be described as a "we-feeling" (Naparstek et al., 1982:19). Sometimes the social group and territorial meanings of community coincide, such as in a Jewish or black ghetto, a Hispanic barrio, a white ethnic neighborhood.

For our purpose we can draw upon these several definitions to describe *a neighborhood community as a people within a limited territory possessing shared values, common interests, and norms of conduct, engaging in social interaction and mutual aid, and having their own groups, associations, and institutions to help meet their basic needs.* The values and norms needn't be absolutely unitary, for a neighborhood can have pluralistic features found in the macro community, but there should be strong enough social bonds to respect differences. The people won't necessarily rely on neighborhood institutions for all their needs and many of their organizations may be quite informal, but they will have various means for collective action.

Let us examine the principal traits of neighborhood communities in greater detail.

FAMILY, FRIENDS, NEIGHBORS

A sizable portion of the neighborhood social fabric is woven from the relationships of family, friends, and neighbors (Tomeh, 1967; Ahlbrandt and Cunningham, 1979:199). Of this threesome, only neighboring is rooted exclusively in the neighborhood, though a large part of primary family life occurs there and the neighborhood is the base for many friendships.

The family particularly is a crucial part of the neighborhood community. Families are not all alike, though. They differ in their composition (one or two parents; whether a third generation or other relatives reside in the same home), in their authority relationships (whether there's one boss or shared authority), and in the relative strength and cohesion of the family unit. Work and commuting patterns have their effects (whether one or both parents work; if the latter, child-care arrangements; hours parents are away from home). The extent that relatives live nearby and how often they get together influence how much time family members have for neighboring and friendships. Background factors of religious affiliation, ethnic and racial group, and social class make a difference.

The family is an important social institution for educating the young, inculcating social values, and instilling behavioral norms. Where families are strong, they are a major source of social control of children and youth. Where they are weak, other social influences are stronger, such as peer groups, school, church, other social institutions, or maybe a criminal element. Adverse societal forces impacting heavily upon a particular neighborhood, such as unemployment and racial discrimination, may contribute to weakening the family and in that manner reinforce other problems confronting the neighborhood. Yet, in many poor neighborhoods, the network of kinship is a major support for human survival (Hannerz, 1969; R.A. Wilson, 1971; Stack 1974; Martineau, 1977).

As noted in the previous chapter, many people choose their homes where families, friends, and other people like themselves live. This may be the neighborhood where they grew up, or a new neighborhood where they make new friends. A study in the Detroit area, for example, found that many of the friends of young men are drawn from childhood and teenage days but as they marry and become fathers, they make new friends in the neighborhood they have settled (Fischer, 1977:85). It takes more than propinquity, though, for friendships are based upon shared values and interests and an emotional bonding. So, friendship patterns jump and skip around the neighborhood, whereas neighboring tends to occur much closer to home. In such a manner, friendships are interlaced throughout a neighborhood, thereby strengthening the social fabric. Friendships, of course, are formed elsewhere—at work, at church, among people sharing particular interests beyond the neighborhood bounds, and this helps create ties to the broader city and metropolitan community.

Neighboring, while less intense than family relationships and friendships, has a unique role in weaving the neighborhood social fabric because it ties together people living next to one another or close by, as we saw in Chapter 2. Sometimes couples and whole families form a neighboring relationship with one another, but often individual family members engage in neighboring separately.

Within the neighborhood social structure, family, friends, and neighbors play different roles. Eugene Litwak and Ivan Szelenyi have hypothesized that "neighbors can best handle immediate emergencies; kin, long term commitments; and friends, heterogeniety" (1969:465).

PEER GROUPS

Geographically based peer groups are another manifestation of community present in neighborhoods. Often quite informal, peer groups consist of persons with similar interests who are usually about the same age. In addition to providing a basis for friendship, they function as a "reference" group, that is, they offer values and norms of conduct to guide their members' attitudes and behavior.

The youngest are children's play groups. They have a life of their own with distinctive folklore, vocabulary, and modes of behavior. Children's groups achieve cultural continuity as they hand along rhymes and games from one generation to another (a span of three or four years) without adult intervention.

Youth groups may or may not be neighborhood-based, for teenagers seek increased mobility and opportunities to get away from their home and immediate neighborhood. Hence, gatherings on shopping center parking lots. But many do have a precise turf. The most disturbing to the general public are inner city youth gangs, which keep springing up in neighborhoods suffering from economic deprivation, social discrimination, and the resulting breakdown of the family structure as a strong instrument for social control (Thrasher, 1927; Cloward and Ohlin, 1960; Suttles, 1968). From their members' perspective, youth gangs represent a functional adaptation to their particular social situation (even though it might be disfunctional for their long run safety and economic advancement). For themselves they form a minicommunity with established norms, shared values, and patterns of mutual aid, no matter how the broader society judges their activities.

Similarly adult corner groups play important roles in the lives of their members. There is still validity to William F. Whyte's description (1943a) of how male groups in Boston's Italian-American North End provided their members social interaction, mutual support, and connections with neighborhood organizations and political networks. In describing low-income, black streetcorner men in Washington, D.C., in the early 1960s, Elliot Liebow emphasized the importance of personal relationships (1967:161-162):

> On the streetcorner, each man has his own network of these personal relationships and each man's network defines for him the members of his personal community. His personal community, then, is not a bounded area but rather a weblike arrangement of man-man and man-woman relationships in which he is selectively attached in a particular way to a definite number of discrete persons. In like fashion, each of these persons has his own personal network.

Drive around any city and you'll see that such corner groups still abound in many lower income neighborhoods.

In the suburbs coffeeklatches of housewives and parties where neighbors get together are other examples of neighborhood-based peer groups. Also, in all sections of the metropolis there are clubs of all sorts where members are drawn from within the neighborhood, sewing circles and craft groups, elderly men gathering in the park to play cards and talk, and lots of other kinds of self-organized activities with regular participants. As with other traits of community, peer groups vary considerably among neighborhoods, but some are present everywhere and add richness to the social fabric.

SOCIAL NETWORKS

Individual patterns of friendships, neighboring, and peer group relationships connect and overlap, forming an intricate social network within the neighborhood and reaching beyond. A social network, according to anthropologist J. Clyde Mitchell (1969:2), is "a specific set of linkages among a defined set of persons, with the additional property that the characteristics of these linkages as a whole may be used to interpret the social behavior of the persons involved." (Also see Barnes, 1954; Bott, 1957; Leinhardt, 1977)

Claude S. Fischer has explained the significance of these social networks in the urban setting. (1977:vii):

> Individuals are linked to their society primarily through relations with other individuals; with kin, friends, coworkers, fellow club members, and so on. We are each the center of a web of social bonds that radiates outward to the people whom we know intimately, those whom we know well, those whom we know casually, and to the wider society beyond. These are our personal *social networks.* Society affects us largely through tugs on the strands of our networks — shaping our attitudes, providing opportunities, making demands on us, and so forth. And it is by tugging at those same strands that we make our individual impacts on society — influencing other people's opinions, obtaining favors from "insiders," forming action groups. Even the most seemingly formal of institutions, such as bureaucracies, are in many ways, to the people who know them well, frames around networks of personal ties. In sum, to understand the individual in society, we need to understand the fine mesh of social relations between the person and society; that is, we must understand social networks.

Folk wisdom summarizes this idea by saying, "it's not what you know but who you know."

While the individual is at the center of her or his personal contacts with neighbors, friends, kin, interest sharers, fellow workers, merchants, service deliverers, and others, she or he is but a single node in the neighborhood network. The individual may die or move away, but the social network persists. A newcomer will move in and establish his or her own individualized pattern, but one connected to the existing, though ever-changing, neighborhood network.

People treasure the web of social relationships found in their neighborhoods, for it meets their need for human contact. These social networks can also be instrumental in helping residents deal with life's problems.

INFORMAL HELPING NETWORKS

Evidence for the importance of informal helping networks to individual health and well-being has been summarized as follows (Froland et al., 1981:21):

> Individual caregivers in an individual's personal network, made up of family, friends, and neighbors, remain the primary reference point for those seeking and obtaining help (Gourash, 1978). Studies conducted over the last several decades show this source of help to be relatively unchanged in its dimensions (Kulka, Veroff & Douvan, 1979). An individual's social network is a major factor in defining the nature of problems, providing help, influencing what sources of outside help will be obtained, and aiding in adjustment to a wide range of acute and chronic

problems (Froland, 1979). Moreover, individual reports suggest that the help received from family, relatives, and neighbors is about as helpful as that received from professionals (Lieberman & Mullan, 1978) and, in some instances, more helpful (Eddy, Paap & Glad, 1970). Informal helping networks are well suited for providing concrete advice, emotional reassurances, an immediate response, long-term caring, and everyday assistance.

In addition to kin, friends, and neighbors, these informal helping networks draw in coworkers, clergy, mutual aid and self-help groups, ethnic and fraternal organizations. Among them, some individuals stand out as natural helpers. They "exist at every socioeconomic level and in every community. Their role does not rely on education, age, sex, or social class. Natural helpers are 'ordinary people' who are seen by others to be especially resourceful and empathetic" (Naparstek et al., 1982:81). (Also see Smith, 1978; Gartner and Reissman, 1980; Gottlieb, 1981.)

In a study of neighborhoods in the Detroit area, Donald I. Warren traced the patterns of helping networks (1981). Interviewers asked people about recent concerns not requiring high expertise for solutions. Examples were feeling blue or tense, thinking about retiring, going back to school, changing jobs, and suspicious people in the neighborhood. As helpers, married persons turned most frequently to their spouse (82 percent). Of working persons 42 percent used coworkers as helpers. Other helpers were friends (41 percent), relatives not living in the same household (37 percent), neighbors (27 percent), and a variety of professional workers, such as police, doctor, counselor, teacher, and clergy (none more than 8 percent). Except for coworkers, most of these helpers were within the neighborhood. When the interviewers asked if the respondents had been helped by a neighbor during a life crisis in the past year (such as personal injury, serious illness, death of a close family member, change of job, crime victimization), 56 percent said they had. Beyond directly helping, neighbors also referred individuals to other helpers, formal and informal. Warren called this the gatekeeper function, a vital part of the helping network.

ORGANIZATIONS AND INSTITUTIONS

In addition to informal helping networks, neighborhoods have various kinds of formal organizations, institutions, and service agencies which assist residents in fulfilling essential functions of life.

Churches and synagogues are key institutions in many neighborhoods. For many European immigrants they formed a linkage between the old and the new. Roman Catholic parishes developed different styles reflecting the dominant group, such as Irish, Polish, Italian, Czech (Greeley, 1977: 92-94). Synagogues gave Jews an institution they could call their own. Likewise in the black community, churches have been important, including independent ones led by charismatic preachers, major denominations (especially Baptist), and small storefront churches. As long as the particular group was kept, or chose to stay, in a specific neighborhood, the church or synagogue was a communal institution of considerable significance. As cities spread out and suburbs grew, churches kept pace with development, and scarcely any neighborhood is devoid of their presence. Sometimes a particular church has become a dominant institution, such as the Episcopal church in eastern, upper class suburbs. Even in neighborhoods where churches and synagogues are found in many varieties and many residents don't attend any, collectively they play an important role in community life.

Schools are also significant neighborhood institutions, especially for families with school-age children but also where they double as community centers for the broader populace. Numerous housing subdivisions have recognized this by according the school a central location, and older neighborhoods often rally around and focus community life upon their school. The public school can be a unifying influence, but the parochial school also contributes to neighborhood life. However, in neighborhoods with relatively few school-age children, schools have less significance.

Voluntary institutions are common in most neighborhoods. They form as people come together to work on some issue, to carry out a specific service, or sometimes just to socialize. Benevolent societies, lodges, recreation associations, buying clubs, housing cooperatives, and community development corporations are some of the forms they take. The neighborhood association is a variety of this species which focuses mainly upon the needs and interests of the territory it serves. Sometimes they spring up spontaneously, but sometimes they require an outside organizer (Taub et al., 1977; also see Chapter 10 on organizing techniques).

Social service organizations and governmental agencies, while not usually indigenous, are also part of the neighborhood institutional structure when they are based there or send out field personnel. As we'll examine in Chapter 12, there are variations on the extent to which residents control these services and facilities, and a major debate exists on whether they should.

Political organizations are often neighborhood-based (called ward or precinct committees). Stores, commercial services, and other businesses are found in most neighborhoods. Later chapters offer a detailed review of the political and economic aspects of neighborhood life.

Not every neighborhood has every variety of organization or agency, and in some places they aren't functioning adequately. But every neighborhood has some kind of institutional structure.

NORMS, VALUES, COMMUNICATION

Less tangible aspects of the neighborhood community are the norms of conduct and the social values held by the residents. Many of these are derived from the broader society or from a subculture which cuts across neighborhood lines. But they are manifested in the neighborhood setting, often conventionally but sometimes in ways deviating from societal norms. Some neighborhoods have considerable pluralism in articulated values and standards of acceptable conduct, while others achieve greater conformity to a unitary set.

Because many neighborhoods are relatively homogeneous, their residents have similar values and postulate a majority behavioral norm. This might be a common expectation of house upkeep, lawn care, use of yards, level and timing of noise (lawnmowers, radios, outdoor parties), and what is acceptable and unacceptable in display of wealth and other status symbols. Parents have rules for their children and teenagers, though the latter might want to observe alternative standards derived from peers and the wider community. The neighborhood norm may place limits on life-styles, at least as visible in the neighborhood, and put pressures for conformity on those who deviate.

Much of the process of communicating neighborhood values and norms occurs informally within the family, neighbor-to-neighbor, or through peer groups. Thus, new residents learn what is expected about putting out the trash, helping one another, control of children and pets. They find out attitudes about cooperating with outsiders, such as police and bill collectors, or shielding their neighbors. Their gossip spreads news and rumors, sometimes quite rapidly.

Neighborhoods also have more formal channels of communication: newsletters, newspapers, church announcements, notices sent home with school children, political handouts, posters, phone trees. The communications process, both formal and informal, is an indispensable part, the nerve system, of the neighborhood community.

DEFENDING THE TURF

Because many people perceive their neighborhood as a place to obtain security and status, the neighborhood social structure functions to achieve these goals. This tendency led Gerald D. Suttles to speak of the "defended neighborhood" (1972). Individual households use such private measures as window bars, burglar alarms, fences, and watchdogs. Apartment buildings have doorkeepers, desk clerks, monitoring systems, and security guards. Face blocks organize to watch over one another's property, the street, and alley. Block associations organize volunteer patrols, and some affluent neighborhoods hire their own security force. Youth gangs guard their turfs, and according to Suttles (1968), this isn't merely a matter of personal belligerence but also an effort to fill a void left by the failure of the police, the courts, and civil adjudicators to assure neighborhood security.

The original developers of some neighborhoods and the residents thereafter used to rely on restrictive covenants in property deeds to control who lived there, with a particular concern for status and other social characteristics. Although such instruments are now unenforceable in courts of law, zoning and market prices are used to place economic restrictions on who can afford to live in the neighborhood. And sometimes residents use coercive methods to prevent persons from what they perceive as an undesirable racial, religious, or ethnic group from moving into their neighborhood, or attending public school there (Buell, 1982).

In short, individual concerns for security and status take on group expressions. This, too, is part of the neighborhood as a social community.

COHESIVE OR DISORGANIZED?

Pulling together the ideas of this and the preceding chapter, we can see that each of us has a personal neighborhood where we have social relationships with family, neighbors, and friends, where we can fulfill some of life's basic functions, where we achieve security and status. How we use our personal neighborhood, and to what extent, varies at different points in our life cycle. As we interact with fellow residents, we create social networks and work out shared values and norms of conduct. Organizations, agencies, and institutions form within the neighborhood to serve our needs. Through these processes, the neighborhood becomes a social community with a life of its own though related to the wider city, and with continuity beyond the time span when any individual may reside within the neighborhood.

Numerous studies have documented many variations of neighborhood community life: white ethnic groups (W.F. Whyte, 1943a; Gans, 1962b; Greeley, 1977); Jewish (Rubin, 1972); black (Hannerz, 1969;

Stack 1974; Hippler, 1974; Martineau, 1977); racial or ethnic mixture (Suttles, 1968); blue collar (Kornblum, 1974); upper class (Baltzell, 1958); suburban (W.H. Whyte, 1956; B. Berger, 1960; Gans, 1967); and many others. These neighborhoods differ considerably in their social structure and achieve varying degrees of success in providing a good life for their residents, but they all function as social communities.

In contrast, another group of social science studies have focused on "slums" and racial ghettos and have described them as places of social disorganization (Thomas and Znaniecki, 1920; McKensie 1923; Wirth, 1928; Zorbaugh, 1929, Drake, 1945; Clark, 1965). They have emphasized social pathologies: adult crime, juvenile delinquency, drug abuse, ill health, bad housing, poverty, breakdown of family life. To be sure, these neighborhoods have a superabundance of problems, many stemming from social and economic forces impacting upon poor neighborhoods. Yet, all of these studies, if read from another perspective, reveal social networks of personal relationships and mutual support, peer groups and indigenous organizations, churches and other social institutions. Some of the ways of life, such as youth gangs and criminal behavior, may be disfunctional in getting on in the wider society, but they nonetheless represent an organized social response to their situation. In this sense, these ways are a manifestation of community. In spite of experiencing a concentration of poverty and other societal ills, these neighborhoods are organized in ways having meaning for their residents.

Any particular neighborhood may be strong or weak in one or more attributes of community, but every neighborhood is a community to some extent — even if its shared values deviate from the dominant norm of the wider society. The relative strength of community is likely to be reinforced if the neighborhood is occupied by people with similar ethnic, racial, or social class background. Nevertheless, a more heterogeneous neighborhood with a broader range of values and life styles can achieve a sense of community through institutional connections and joint action on common problems, or, it can be disjointed and wracked with intergroup conflict. Neighborhoods of a city could be scaled from cohesive to fragmented in their sense of community, but all neighborhoods can be considered social communities because of the presence of social networks, peer groups, associations and institutions functioning within the neighborhood territory.

Exercises

(1) See how many definitions of community you can find. How do they differ from definitions of neighborhood you found in Chapter 1?
(2) Put these definitions of community on a matrix according to key traits. Notice which traits are emphasized and omitted by different definitions.
(3) Write your own definition of community.
(4) Draw a chart of some of the social networks you are a part of, starting with one-to-one relationships you specified in the exercise for Chapter 2 and indicating how these persons are connected to other people.
(5) Which networks occur mainly within your neighborhood? Which ones are connected with the wider city and metropolis?
(6) Describe the principal organizations and institutions you are in contact with in your neighborhood. What functions of life do they help you fulfill?

CHAPTER 4

A PHYSICAL PLACE

As a limited territory within a larger settlement, a neighborhood is a physical place. In providing shelter a neighborhood is composed of dwellings of various types, sizes, and designs. It has buildings for other purposes, such as schools, churches, medical offices, perhaps commercial and industrial structures. There are streets, sidewalks, playgrounds, parks, and other open spaces. As a place, a neighborhood has boundaries, sometimes clearly demarcated and agreed upon but in other cases less distinct and a matter of varied opinion. Yet no matter how precisely bounded and how well designed, the territory becomes a neighborhood only through occupancy and use by its residents. Therefore, this chapter weaves in the social as it considers the physical manifestations of neighborhood.

HISTORICAL DEVELOPMENT

Cities have always had neighborhoods, that is, residential quarters set apart from other major functions — from the central market, main temple or church, palace, fortress.

In *The City in History* (1961), Lewis Mumford described the presence of neighborhoods in different eras. Of Mesopotamian cities of 2,000 B.C. he wrote (p. 74), "Beyond the walled but spacious temple precinct, spread a series of more or less coherent neighborhoods in which smaller shrines and temples serve for the householder." The ancient Greeks gave geometric order to their cities, and this had the effect of "dividing the city into definite neighborhoods, or at least giving that definition visible boundary lines (p. 193)." Roman planning did likewise, creating neighborhoods with their own minor centers and markets (p. 211).

Medieval cities relied less on a formal, geometric plan, and growth was more organic. According to Mumford (1961:310):

> In a sense, the medieval city was a congeries of little cities, each with a certain degree of autonomy and self-sufficiency, each formed so naturally out of common needs and purposes that it only enriched and supplemented the whole. The division of the town into quarters, each with its church or churches, often with a local provision market, always with its own local water supply, a well or a fountain, was a characteristic feature; but as the town grew, the quarters might become sixths, or even

smaller fractions of the whole, without dissolving into the mass. Often, as in Venice, the neighborhood unit would be identified with the parish and get its name from the parish church: a division that remains to this day.

The advent of capitalism, industrialization, and rapid growth of cities brought major changes in patterns of urban development. "Thus the city, from the beginning of the nineteenth century on," Mumford stated (1961:426), "was treated not as a public institution, but a private commercial venture to be carved up in any fashion that might increase the turnover and further rise in land values." The fundamental unit was no longer the neighborhood or precinct but rather the individual building lot. The gridiron pattern with its regularity was tailor-made for surveyor, land owner, real estate speculator, lawyer, and commercial builder. Topography was ignored, and specific functions weren't taken into consideration. "Instead the only function considered was the progressive intensification of use, for the purpose of meeting expanding business needs and raising land values" (p. 424).

Nevertheless, as Mumford observed, people developed neighborhood identity in the growing cities. "Even in the undifferentiated rectangular plan of Manhattan," Mumford observed (1968:58), "a plan contrived as if for the purpose of preventing neighborhoods from coming into existence, distinctive entities, like Yorkville, Chelsea, and Greenwich Village, nevertheless have developed, though they lack any architectural character, except that conditioned by the successive dates of their building."

IDENTITY, FAMILIARITY, SIMILARITY

Manhattan was not alone in gaining neighborhoods in spite of the speculative nature of land development and the relentless application of the gridiron plan in numerous cities. Even though few cities consciously planned neighborhoods, they occurred everywhere for several reasons.

First and foremost is personal feelings we all have for our homes as space we occupy. In spite of what social critics say about the anonymity of big apartment buildings, mass-produced housing, and ticky-tacky suburban tracts, each of us gives our own individual touch to the dwelling unit we live in (even if we move around quite a bit): our furniture, rugs, curtains, pictures, knickknacks, plants. Likewise where possible we personalize the exterior, such as through color choice for the house, doors, and shutters; planting of garden, shrubs, and lawn; arrangement of patio, porch, or balcony (Greenbaum and Greenbaum, 1981).

It's our space, and we identify with it. As Marc Fried has indicated (1963:156):

> In fact, we might say that *a sense of spatial identity* is fundamental to human functioning. It represents a phenomenal or ideational integration of important experiences concerning environmental arrangements and contacts in relation to the individual's conception of his own body in space. It is based on spatial memories, spatial imagery, the spatial framework of current activity, and the implicit spatial components of ideals and aspirations.

We extend this sense of spatial identity from our home to the immediate neighborhood, though this is stronger for some than others. This combines with the natural impulse for neighboring, which we discussed in Chapter 2. Thus, Fried and Peggy Gleicher found both neighboring and spacial identity highly significant to the majority of residents who felt satisfied with an area planners considered to be a slum. They noted (1961:315).

> On the one hand, the residential area is the region in which a vast and interlocking set of social networks is localized. And, on the other, the physical area has considerable meaning as an extension of home, in which various parts are delineated and structured on the basis of a sense of belonging.

Our spatial identity is reinforced by our tendency to spend most of our time in familiar settings. As John T. Dean has explained (1958):

> Most people in patterning their daily activities develop a sort of "beaten path" that they tread from home to work, then back to home, then to lodge meeting, back to home; on Sundays to church and back, and then perhaps a visit to close relatives. Then the pattern repeats. Once developed most persons stay on their "beaten path" and only once in a while get off into the forest where all the rest of the hustle and bustle of what we call "urban mass society" takes place. Each person's "beaten path" is different from all the rest and there are all manner of variety in the range and types of beaten paths. But for the usual person, the beaten path is a narrow walk of life that exposes him to two or three, maybe four, different social environments — that's all.

One of these social environments is the personal neighborhood where we engage in neighboring and use close-by facilities to fulfill some of the basic functions of living. Each of us has a perception of this distinct local environment, differentiated from the rest of the city. It is space we know, where we feel secure, whereas many other parts of the city away from our beaten paths are unfamiliar and sometimes hostile territory, so we perceive (Suttles, 1972:22).

The development process itself has factors making neighborhoods likely to come about. Carried out mainly through the private initiatives of land owners, builders, realtors, bankers, and lawyers, development has occurred mostly on tracts or subdivisions with dwellings similar in style and sale or rental price. Even in the days before zoning, there was a pattern of fairly uniform density and housing types in particular sections. And there was a tendency toward social class differentiation as upper-income housing concentrated in one or a few sections, usually outlying after the first period of settlement, working class housing was situated near industrial areas, and middle-income housing in between. This has created residential sections occupied by people with many similarities, such as occupation, income, ethnicity, race, religion (Hoover and Vernon, 1959). Residents have many of the same values and interests. Thus, the physical similarity of the housing provides the foundation for social similarity. This encourages and facilitates common neighborhood activities.

The need for access to dwellings places houses and apartment buildings on streets, or on off-street courtyards and driveways. Whether gridiron or curved to fit topography, the street, or an alternative accessway, is one of the most important physical features of cities in providing a setting for neighborly contacts and localized activities. We'll examine this closer in a moment, but first we should look at the affects of housing design.

HOUSING DESIGN

In rededicating the British House of Parliament after its reconstruction following World War II, Winston Churchill remarked, "We shape our houses, and our houses shape us." He was referring to the influence of a long, narrow hall of facing seats in the House of Commons, which was best suited for a two-party division. His thought also applies to the effects of design of housing and the space around it upon the creation of neighborhoods.

Housing comes in many shapes and sizes: single-family detached, row (town) houses, duplexes, triple deckers, garden apartments, trailer courts, six-story walkups, elevator apartment buildings, projects with a number of high-rise buildings. Some neighborhoods have mainly one type, others greater variety. Interrelated are the rent level and sales price, strongly influencing which socioeconomic group occupies the housing, and also such factors as age of residents, family types, ethnic or racial group, and life-style. For, people with these varied backgrounds use space differently.

Density, that is, the number of dwelling units per acre, is another influence. Extremely high density sometimes has the effect of drawing people inward to their private space. Extremely low density (houses on lots of two or more acres) provides very few neighbors to relate to. To the extent that people have plenty of habitable space inside, they are less likely to spill out onto the sidewalks and streets. How much of the land is occupied by buildings determines how much is left for outdoor space. The location of such ground space in relationship to the dwelling units is important.

The interrelationship between space within and outside the dwelling units is particularly important. Oscar Newman (1972) has suggested that such space can be arranged in a fourfold hierarchy from private to public.

Private space is found within the dwelling unit, balcony, and enclosed yard. The more space for entertaining and the more facilities within the unit, such as laundry equipment, the less will occupants need to go outside for particular needs. How the yard flows into public space affects neighboring relationships.

Semi-private space includes the front porch, stoop, and unfenced front yard of houses. In duplexes and triple deckers it is the stairway and entrance hallway. In apartment buildings it is hallway space serving a cluster of apartments of the same floor. From these spaces the occupants have contact with neighbors and passersby.

Semi-public space is the lobby and circulation hallways within apartment buildings and courtyards of garden apartments. These spaces are intended primarily to serve the occupants and their visitors, but others can gain access.

Public spaces include undemarcated grounds of apartment projects and streets and sidewalks adjacent to all kinds of housing. The latter are usually owned by local government and are legally public, and, while the former are technically private, unless fenced and guarded they are readily accessible to anyone who wants to enter.

In his concern for defensible space, Newman (1972:4) indicated that architects can create a clear understanding of the function of a space and who its users are and ought to be by (1) grouping dwelling units to reinforce associations of mutual benefits, (2) delineating paths of movements, (3) defining areas of activity for particular users through their juxtaposition with internal living areas, and (4) by providing for natural opportunities for visual surveillance. The reciprocal of security is neighborliness, so many of these same design features can create a setting where neighboring is more likely to occur and neighborhood activities are more likely to develop. Indeed, in a subsequent study, Newman (1979) found that the greater the size of apartment buildings (at least for federally-assisted housing), the less use residents made of public areas in their development, the less social interaction they had with their neighbors, and the less sense of control they felt for interior and exterior public spaces.

Much earlier anthropologist Anthony F.C. Wallace compared livability of a low-rise public housing project in Philadelphia with a high-rise public housing project in New York (1952). He found the physical design of the low-rise project relatively satisfactory for the low-income families living there. Tenants liked the dwelling units. There were private (or potentially private) yards and gardens. Large families had row houses. Children could play near home within sight of their mothers. In contrast, there were many crises built into the high-rise project, such as limitations of contact between mothers and children, absence of any private outdoor space, necessity for management to play a large role in controlling open land spaces, halls, and elevators, and the consequent lessening of responsibilities for the family and tenant organizations.

The experience of housing authorities who ignored these findings and built large, high-rise public housing projects for low-income families proved that design had an effect, in this case negative, upon creating a sense of neighborhood, a positive community identity, and a safe environment. Long, unguarded hallways, elevators, and stairwells and public spaces where no one was responsible, combined with eligibility rules which moved out the abler persons whose income rose while concentrating families with numerous problems, proved to be disastrous (Rainwater, 1970; Yancey, 1971). Moreover, federal regulations prohibited inclusion of stores and commercial services, which are needed in a well-rounded community. A number of these buildings were demolished within 20 years of their construction, and many more have been virtually abandoned and boarded up.

There is no guarantee that good design and defensible space will produce a safe and cohesive community, for a strong social fabric is also required (Merry, 1981b). However, a well-articulated physical framework can create opportunities where informal neighboring and organized activities are more likely to occur.

NEIGHBORHOOD SPACE

Residents use many neighborhood spaces which aren't part of their dwellings, and this has an influence on community life. As defined by Randolph T. Hester, Jr. (1975:20):

> Neighborhood space is that territory close to home, including houses, churches, businesses, and parks, which, because of the residents' collective responsibility, familiar association, and frequent shared use, is considered to be their "own."
>
> Included in this definition are such spaces as parks, street corners, storefronts, alleys, rooftops, sidewalks, schoolyards, playgrounds, parking lots, streets, paths, junkyards, front porches, streams, abandoned lots, secret niches, plazas, churchyards, trash dumps, woods, bus stops, front steps, gardens, outdoor cafes, phone booths, forbidden places, favorite places, utility spaces, floodplains, ponds, greenways, conservation easements, beautification areas, and transportation corridor spaces.

The street is the most common neighborhood space, and in many respects the most important. It serves as a passage for people and vehicles — in, out, and through the neighborhood. Enlarged to encompass all the space between facing houses and apartment buildings, it is also a locale for social interaction, recreational activities, informal communication, and sometimes for the purchase of goods and services. As with other aspects of neighborhoods, there is a considerable range in the ways and intensity residents make use of their streets. But rarely is the neighborhood street only an impersonal passageway, no matter how it may seem to an outsider.

During the past 50 years a variety of studies have shown the importance of the street and accessways. In the previous chapter, we cited reports about how in certain neighborhoods the street corner serves as a gathering place for peer groups, especially males. There they have fellowship and tie into networks of social interaction and communication with the wider neighborhood and beyond.

Studying Boston's West End in the 1950s, Herbert J. Gans (1962) found that street life was important for the entire population. In fact, he observed that residents didn't think of the West End as a single neighborhood but rather related mainly to the street where they lived, the stores they frequented, and their particular ethnic group (Italian-American was the largest with smaller numbers of Jewish, Polish, Irish, Albanian, Ukrainian, and Greek households). Most of the area's communal life took place on the street: mostly women and children during weekdays, though with some older, retired men, and some younger and middle-aged ones who were unemployed, worked night shifts, or made a living as gamblers. On Sunday morning the streets filled with people visiting neighbors and friends before and after church. Gerald D. Suttles (1968) recorded similar activities in the 1960s in the Addams area of Chicago's Near West Side, occupied by Italians, blacks, Puerto Ricans, and Mexicans. He particularly stressed that street life provided a vital link in the communications network and governed much of what the residents knew of one another beyond the range of personal acquaintance.

Jane Jacobs has emphasized another important function of neighborhood streets and sidewalks: personal safety. "The first thing to understand," she noted (1961:31-32), "is that the public peace — the sidewalk and street peace — of cities is not kept primarily by the police, necessary as police are. It is kept primarily by an intricate almost unconscious, network of voluntary controls and standards among the people themselves, and enforced by the people themselves." For a street to be safe, there must be eyes upon it, eyes belonging to the "natural proprietors": shopkeepers, people on errands, residents watching from windows. Enterprises open in the evening are especially important, such as stores, bars, and restaurants. Activities attract people, and this increases the number of eyes on the street. A well-used street is apt to be a safe street. (For the impact of vehicular traffic, see Appleyard, 1981.)

Even where street life is less intense, streets, apartment courtyards, and connecting walkways play an important role in setting the framework for personal contact and neighboring. Thus, two separate studies of married couples housing at college campuses after World War II discovered that, among residents who had many similarities in current status and pursuits, the placement of apartment units along courtyards and passageways was the greatest determinant of neighboring and friendship patterns (Caplow and Forman, 1950; Festinger et al., 1950).

Studies in postwar suburban developments revealed similar patterns. Thus, William H. Whyte, Jr. (1956) noted that in Park Forest, Illinois, where garden apartments clustered around parking bays, these courts became the primary cell of social life. In Park Forest's owner-occupied, single family dwellings and also in similar housing in Levittown, New Jersey studied by Gans (1967), small social groups formed along and across streets but rarely through backyard connections with the other side of the block. Children played on walkways along the street and in courtyards, and this affected the neighboring relationships of their parents, especially nonworking mothers. Social groups along streets and within courtyards limited their size, rarely exceeding a dozen couples, about the maximum which could fit in the living-dining area of a host couple's home. But as Gans pointed out (1967:154), "Propinquity may initiate social contact but it does not determine friendship." Friendship formation depended upon similarities, which the nature of the subdivision produced, and social compatibility, which represented sorting out of likes and dislikes. But contact along the street and courtyard set the stage. (Also see Mayo, 1979.)

Randolph Hester has noted that lower income residents make greater use of space in front of the house (porches, steps, sidewalks, street) than upper income residents, who more often rely on the controlled setting of private yards (1975:32). Gloria Levitas (1978:231-232) offered a similar finding, stating that "the role of the street and the nature and content of social interaction vary with class, ethnic group, age structure, and type of specialization of the neighborhood." She stated that increasing specialization and compartmentalization of society have removed indoors many of the socially cohesive activities once found in the street, especially entertainment, marketing, information, and personal services. In some suburban developments there is no sidewalk, and in central areas with high-rise apartment buildings, firm boundaries separate building and street. "Only in the slum and in the dwindling ethnic enclaves and blue collar areas does the street still seem to function as a locus for public life."

Hester, though, believes that there has been significant general increase in the use of neighborhood open space for leisure activities, and also for work space and political gathering space. But what he calls work isn't as intensive as the marketing and personal services which Levitas says have disappeared. But Hester seems to be right, at least to the extent that organized city neighborhoods are increasingly using outdoor space for such communal events as block parties, street fairs, cleanup and beautification campaigns (see Chapter 11, p. 167). But there is considerable variation among neighborhoods, with the least street life occurring in high density, upper income areas.

NEIGHBORHOOD PLANNING UNIT

In making use of streets and other spaces, residents take what's there, regardless of intent of original design, or lack of intent. In that way they have made neighborhoods out of real estate developments, even those cast in the gridiron pattern.

But not all new development has been planless and unconcerned with neighborhood amenities. From the late nineteenth century onward, there have been new subdivisions where developers and architects have consciously tried to produce a superior plan with streets curved to fit topography, houses carefully sited on lots, parks and schools designed into the plan. One of the first was Riverside, a 1,600-acre tract near Chicago, laid out by Frederick Law Olmsted and Calvert Vaux in 1869. Other suburban subdivisions of notable design undertaken by profit-oriented developers were Garden City, Long Island (also begun in 1869), Roland Park, Baltimore (1891), Country Club District, Kansas City, Missouri (1905), Forest Hills, Long Island (1913), River Oaks, Houston (1923), and Palos Verdes, Los Angeles (1923). As Arthur B. Gallion and Simon Eisner have pointed out (1975:130): "Each of these developments marks a high level of planning in living environment. There was no serious intention on the part of subdividers to cope with housing for low income people in these communities. They were intended for the upper income group and were promoted accordingly."

During this same era theorists were developing ideas which eventually led to articulation of principles of planned neighborhoods. Particularly influential was a book by Ebenezer Howard of Great Britain, published in 1898 as *Tomorrow: A Peaceful Path to Land Reform* and republished in 1902 as *Garden Cities of Tomorrow* (1965). He proposed what are now called satellite cities, separated by a greenbelt from the older metropolis. Others shared his conviction that a different pattern of urban development was necessary and possible, and an international town and country planning movement grew.

The neighborhood aspects received their clearest expression in ideas offered by Clarence Perry, culminating in a chapter he wrote for the monumental *Regional Plan of New York and Its Environs* (1929), sponsored by the Russell Sage Foundation. He advocated that the building block of city planning should be the neighborhood unit, designed to carry out the following principles (1929:34-35):

(1) *Size:* A residential unit development should provide housing for that population for which one elementary school is ordinarily required, its actual area depending upon population density.

(2) *Boundaries:* The unit should be bounded on all sides by arterial streets, sufficiently wide to facilitate its by-passing by all through traffic.

(3) *Open spaces:* A system of small parks and recreation spaces, planned to meet the needs of the particular neighborhood, should be provided.

(4) *Institution sites:* Sites for the school and other institutions having service spheres coinciding with the limits of the unit should be suitably grouped about a central point, or common.

(5) *Local shops:* One or more shopping districts, adequate for the population to be served, should be laid out in the circumference of the unit, preferably at traffic junctions and adjacent to similar districts of adjoining neighborhoods.

(6) *Internal street system:* The unit should be provided with a special street system, each highway being proportioned to its probable traffic load, and the street net as a whole being designed to facilitate circulation within the unit to discourage its use by through traffic.

Partly through the influence of this report and partly through parallel initiatives of architects and planners, these ideas were put into practice during the 1930s in such developments as Sunnyside, Long Island; Radburn, New Jersey; Chatham Village, Pennsylvania; Mariemont, Ohio; and three greenbelt towns constructed as a New Deal program in Maryland, Ohio, and Wisconsin (Dahir, 1947). Typically the school and playfield were centrally located, through traffic was routed around the development, residential streets were laid out to fit land contours and add interest to design, and often separate walkways and bicycle paths were constructed.

However, there was some dissent from the idea of the neighborhood planning unit, notably by a Chicago planner, Reginald Isaacs (1948a), who sided with Louis Wirth's views on urbanism (see p. 30), arguing that "in the city people become members of groups larger than neighborhoods and merely *reside* in residential areas in contrast to *living* in rural or village neighborhoods as was true in the past" (1948a:18). Moreover, he pointed out that the neighborhood was an instrument of racial and social class segregation, aided and abetted by realtors, lending institutions, and the Federal Housing Administration (FHA), whose underwriting manuals were directed toward keeping out "inharmonious" and "incompatible groups," which in practice meant blacks, Jews, and white ethnic groups.

In rebuttal, Lewis Mumford insisted that practices of segregation by race, caste, or income had nothing whatever to do with the neighborhood principle. Instead he advocated that a neighborhood "should, as far as possible, be an adequate and representative sample of the whole. . . . The mixture of social and economic classes within a neighborhood should have its correlate in a mixture of housing types and densities of population" (1968:75). Catherine Bauer also advocated heterogeneous neighborhoods (1945). (Also see Churchill, 1945; Herbert, 1963.)

Notwithstanding this debate, in the years since World War II numerous suburban developments have put into practice the physical features of planned neighborhoods, as you can observe in flying over any metropolis in the United States. You see looped streets and sometimes houses and apartments clustered on cul-de-sacs with common open space in the rear. As in Perry's original articulation, many of the new subdivisions limit commercial activities to the periphery, but some have provided for convenience stores and other commercial uses in central locations along with schools, churches, and other community facilities. Some of the larger developments have grouped several neighborhoods around a large shopping center. Socially, though, most new subdivisions are occupied by a fairly narrow socioeconomic stratum.

In older parts of cities undergoing redevelopment, the neighborhood unit concept has competed with a tendency of many architects to design single projects without reference to surroundings, city planners' notion of land use separated through zoning, and federal regulations which have kept commercial services out of housing projects. Jane Jacobs excoriated the product of this kind of thinking in *The Death and Life of Great American Cities* (1961), a book pleading for neighborhood diversity.

ESTABLISHING NEIGHBORHOOD BOUNDARIES

If a neighborhood is a physical place with limited territory and the focus of neighborhood planning, it must have neighborhood boundaries. Or at least that is what many planners, government officials, neighborhood leaders, and social scientists believe. This raises questions of how large is a neighborhood, how do you decide upon its boundaries, and who makes this determination. These issues go back to one's basic definition of neighborhood.

Each profession brings its own interests and concepts to the task of drawing up neighborhood boundaries. City planners are interested mainly in physical aspects of neighborhoods, so they look for such recognizable features as rivers and streams, railroads, interstate highways, and major streets. Realtors and mortgage lenders like to group houses of similar type and sales value, and realtors will sometimes extend the bounds of a prestigious neighborhood for marketing purposes. Municipal administrators conceive of service areas, school officials envision school attendance areas, and politicians think of voting precincts. Sociologists search for homogeneity of social classes, ethnic and racial groups. Sometimes these outsiders will also look for historical factors and perhaps self-identity as indicated by existence of neighborhood associations.

Residents have a different perspective. For example, surveyors from the Joint Center for Urban Studies of MIT and Harvard University, asked people in Houston, Dayton, and Rochester (Birch, 1979:36), "What do you think of as the boundary or borders of your neighborhood — where does it begin and end?" Single-family occupants spoke of an immediate neighborhood of "the houses around my house," or the block. Apartment dwellers perceived their immediate neighborhood as their floor, wing, building, or apartment complex but not the surrounding houses. Single family dwellers next considered a larger homogeneous neighborhood extending to where the market value of housing noticeably changes, or if heterogeneous, an area bearing a name, the school district boundary, the civic association area, or area demarcated by a major traffic artery. Both apartment dwellers and single-family occupants excluded any area where they wouldn't want to live, no matter how close. Both also spoke of a larger district — a section of the city or suburbs.

In many cities these larger districts are served by community newspapers. In 1952 when he wrote about the community press in Chicago, Morris Janowitz coined the phrase "community of limited liability" to describe the district to which residents have some attachment but a limited amount of social and psychological investment (1967:211). Twenty years later Gerald D. Suttles and Albert Hunter suggested that this was the third level of a hierarchy which begins with the "face-block" and expands to the "defended neighborhood" and then goes beyond to a fourth area, an "expanded community of limited liability" encompassing an even larger geographic district of the city (Suttles, 1972:54-64). I agree with the idea that people perceive and use several types of subareas, arranged in a size hierarchy. The terminology I prefer is (1) immediate personal neighborhood, or face block, (2) functional neighborhood, and (3) community district. The second level is the most common focus of neighborhood associations and service delivery arrangements.

In Pittsburgh the Neighborhood Atlas project worked with residents to draw up a citywide map of nonoverlapping neighborhoods of this second level. They found people somewhat egocentric in their definitions, placing their own homes near the center of the neighborhood they described even though it might be near the border of the territory encompassed by the local neighborhood association. However, personnel from the Atlas project were able to sort out the boundaries, and by making compromises and assigning areas not claimed by any neighborhood, they were able to produce a citywide map of 78 distinct neighborhoods in this city of 425,000 people.

In other cities, planning or community development departments have drawn the boundaries, relying heavily upon physical features and usually trying to include whole census tracts so as to have data for analysis. (For the 1980 census, though, the U.S. Bureau of the Census has provided data for a city's own defined neighborhood areas.) In some cities, such as Eugene, Oregon, neighborhood associations draw up their own boundaries, subject to approval by the planning commission or city council, which eliminates overlapping and assigns unwanted areas. Portland, Oregon has a recognition process which permits neighborhood associations to have overlapping boundaries in realization that neighbors sometimes have different perceptions of the area they identify with.

As to size, usually residents define smaller neighborhoods than outsiders. Clarence Perry conceived the neighborhood unit as an area served by an elementary school, that is, perhaps 6,000 to 7,000 people (1929). Donald I. Warren also used the elementary school area as his working definition of neighborhood (1981). City planning departments look for areas ranging from 4,000 to 12,000, and sometimes larger. In cities with officially defined neighborhoods, they show a considerable range in size, reflecting varying configurations around the city. In these situations, some neighborhoods might hold 20,000 to 30,000 people while some 1,000 or less.

Among city officials, neighborhood size tends to be correlated with city size, for they try to limit the number of neighborhoods to a manageable number. Thus, over the years Chicago planners have kept track of 77 community areas, derived from 75 identified defined by University of Chicago sociologists in the 1920s, though when Albert Hunter (1974) interviewed residents in all parts of the city in 1967-1968, he identified 206 smaller neighborhoods. And when Ed Marciniak (1981) studied one of the community areas, he determined that it contained two distinct communities, one of them with 13 individual neighborhoods. In New York the City Planning Department has divided the city into 59 community districts but recognizes that each of them contains many separate neighborhoods. Thomas Broden and associates analyzed 29 cities which planners and administrators had divided completely into neighborhoods. They found that 14 defined what I have called "functional" neighborhoods (the second level, larger than personal neighborhoods), 7 used larger units (multineighborhood, community districts), and 8 combined these kinds of areas into two tiers (1979).

This gets us back to our earlier attempt in Chapter 1 to define "neighborhood." It depends upon your perspective and purpose. There are indeed neighborhoods. They are physical places as well as social communities. But because urban residents are mobile and carry out their lives in different ways, they inevitably have different perceptions of the neighborhood place they relate to. Therefore, any effort to divide a city into completely nonoverlapping neighborhoods has to contain compromises and arbitrary decisions.

Exercises

(1) Draw a map of beaten paths you follow regularly. Indicate important places and other features.

(2) Draw the boundaries of geographic subareas of the city or suburbs important to you. If more than one, are they separate or do they overlap in increasing size?

(3) Make a similar map for earlier periods in your life when you lived in a different neighborhood or used your neighborhood differently.

(4) Describe your activities in public, semi-public, and semi-private spaces near your dwelling unit. Who are you involved with there?

(5) Examine two or three different kinds of neighborhoods and describe activities occurring in spaces outside the dwelling unit.

(6) What is the more important influence on these activities: (a) housing design, site plan, and layout of community space or (b) the social and subcultural setting? If both are important, how do they interrelate?

CHAPTER 5

A POLITICAL COMMUNITY

A territorially based community — of which neighborhood is one variety — has shared values, common interests, and accepted norms of conduct. To function, the community requires instruments to get things done and to maintain orderly relationships among its members. One such instrument, though not the exclusive one, is government, distinguished by its capacity to exercise power and authority over members of the community. When people relate to government as citizens, give it authority to act, are subject to it, pay taxes, and receive services, they constitute a political community.

Neighborhoods function as political communities in several ways. The most common is by serving as a base for dealing with governmental jurisdictions of broader domain: city, county, school district, other special districts, state, and national. All told, these jurisdictions are responsible for the preponderance of governmental programs and processes serving and impacting upon neighborhoods. Much of what they do is centralized at city hall, the county courthouse, the state capitol, and agency headquarters, but a considerable amount is decentralized to facilities and field services functioning within neighborhoods. As a political community, a neighborhood relates to the centralized and decentralized manifestations of broader domains by acting as an electoral constituency and an interest group. In the latter capacity, the neighborhood also deals with corporate and institutional interests in the private sector.

Less frequently neighborhoods have their own general purpose governments, most notably small suburban municipalities of neighborhood size. But some city neighborhoods have moved toward greater self-governance through the formation of neighborhood councils having policy roles and nonprofit corporations handling service delivery and initiating physical and economic development.

The combination of the internal and external governing processes and the relationships between them defines the scope of the neighborhood as a political community.

Dealing With Broader Domains

To be just, government must derive its powers from the consent of the governed. So says the Declaration of Independence. This keeps government under the control of the people and gives it legitimacy. As Seymour Lipset has written, "Legitimacy involves the capacity of the system to engender and maintain the belief that the existing political institutions are the most appropriate ones for the society" (1963:66).

Historically in the United States we have experienced an increasing variety of methods for gaining consent: representative government, measures of direct democracy, voting with expanding eligibility, political parties, interest groups, and structured citizen participation processes. Neighborhoods, functioning as political units, get involved in all of these.

VOTING

Citizens in every neighborhood in the United States have an opportunity to vote in elections for president and vice president, members of Congress, governor, and state legislators. Depending upon neighborhood location and structure of local government, they might also vote for candidates for mayor, county executive, city council, county board or commission, town or village offices, school board, and a variety of other local offices. Some cities also have neighborhood councils (known by a variety of names), chosen by the residents, and a few locales have elected community school boards.

Neighborhoods have varying degree of voter participation, and this affects their influence upon election outcome. Historically population segments with less than average voting include blacks, Hispanics, persons with lesser education, blue collar workers, the unemployed, low income persons, renters, and persons who have lived a short time at their present residence (Verba and Nie, 1972). However, as Wolfinger and Rosenstone have determined (1980), length of education makes the greatest difference, though this effect is lessened as individuals grow older and their practical experience increases their interest in and knowledge about electoral issues and candidates.

Because neighborhoods tend to be stratified by income, occupation, education, race, and ethnicity, any given neighborhood is composed mostly of a limited stratum. Where a neighborhood has a sizable population group traditionally with a low voter participation, that neighborhood has less electoral impact. However, intensive voter mobilization can produce higher than average voter turnout. As

Milbraith and Goel have observed (1977:138), "Persons are more likely to turn out for elections they perceive to be important."

REPRESENTATION

Voter participation makes a difference because elected officials are aware of where their support has come from. As a general rule, the smaller the area of representation the greater the likelihood that candidates and elected officials will pay attention to neighborhood constituencies. Members of elected neighborhood boards are clearly accountable to the residents. In small cities with district elections, a council member might serve a single neighborhood, and the same in medium-size cities with a large council. But usually in larger cities, council districts encompass several neighborhoods and so do state legislative districts, yet not too many, so that neighborhoods can have an impact. In contrast, where city councils are elected at large and also in mayoralty elections, a particular neighborhood is but one voice among many and therefore has to be assertive to be heard. In this situation, neighborhoods are probably most influential where they also constitute another kind of voting bloc, such as black, Hispanic, an ethnic group, or where they are well-connected through party politics or ties with holders of economic power.

Neighborhoods use their votes and their continuing relationships with elected officials in a variety of ways to secure benefits from and prevent adverse actions by governments of wider domain. (For more on relationships with council members, see Davidson, 1979. On the issue of district elections, see Karnig, 1976; L. Cole, 1976; Robinson and Dye, 1978; Taebel, 1978; Herlig and Mundt, 1982.)

PARTY POLITICS

In American democracy political parties are an integral part of electoral and governing processes. Usually they are organized on a foundation of voting precincts and wards. In some places, though not everywhere, wards serve as the basis for city council districts, but precincts are always smaller. Typically a neighborhood has a number of precincts. It might approximate a single ward, contain several complete wards, or fall into parts of different wards.

Historically — particularly in immigrant neighborhoods — the precinct committeeman or ward chairman was an important figure, a dispenser of benefits, a mediator with public agencies, and a friend. The political committee was a kind of neighborhood association. In some places, these political workers themselves benefited economically through bribes, payoffs, and inside information. Public reaction to this corruption led to institutional changes (such as at-large and nonpartisan elections), which diminished their role. Moreover, with the growth of public programs, bureaucracies took over much of their social service activities. Nevertheless, in cities where political parties are strong, precinct and ward committees and their officers remain as significant actors in the neighborhood political community. This is more likely to occur where local elections are partisan, that is, where the ballot identifies candidates by political parties.

INTEREST GROUPS

Another way the neighborhood political community functions is to act as an interest group in dealing with broader governmental domains, institutions, corporations, and other kinds of interests. An interest group consists of people who share a common concern and who work together to promote their shared interest. They are usually known by the name of their common interest (such as business, labor, agriculture, a cause) or a population segment (such as women, blacks, the aged). As such, interest groups have long been considered an important factor in American politics and governance (Truman, 1951; J. Wilson, 1973), and this carries over to neighborhoods (O'Brien, 1975; Henig, 1982).

Sometimes residents act together on a temporary, ad hoc basis to protest or to promote their interests as a neighborhood. But many neighborhoods have realized that to be effective over the long haul, they need some kind of ongoing organization. This takes a variety of forms and expressions.

Protection. As noted in Chapter 2, we as individuals look to our neighborhood to offer security and status. Because many forces — social, political, economic, environmental — impact the neighborhood, we realize that we alone cannot keep it what we want it to be. Therefore, we get together with our neighbors to protect the values we cherish, our properties, and our personal safety.

This may take the form of a homeowners association, comprising only residents owning property. It may be a more inclusive neighborhood association, but still oriented toward protective measures. Block residents get together to help each other secure the safety of person and possessions. Condominium associations and housing cooperatives are also concerned about safety. Some youth gangs have protection of turf as a major mission (though others are more concerned about economic gain from illegal activities).

Although these kinds of organizations aren't recognized as governments, they focus upon one of the functions which governments carry out. In our philosophical lineage, this was expressed in the eighteenth century by John Locke, who believed that people unite in political society and form government "for the mutual preservation of their lives, liberties, and estates" (1960:395). Thus, protection is an expression of the neighborhood political community.

Advocacy. To protect person and property and to achieve other aims, neighborhood associations reach out and act as advocates in dealing with the city, the county, state and federal agencies, private businesses, nonprofit institutions, and anybody else who has impact upon neighborhood conditions. In doing so, they deal with a wide range of issues, such as rezoning, police protection, community development plans, highway location, quality and quantity of public services, lending practices of financial institutions, and many more.

In advocacy, neighborhood organizations don't wait to be asked for their opinion. They take the initiative. They use a variety of tactics, ranging from quiet contacts to noisy protest. Sometimes they work through electoral/political party processes, but they often use other avenues to gain access to decision makers and to press for policies favorable to the neighborhood. As we'll explore in Part II, neighborhood advocacy is a major focus of community organizing efforts.

STRUCTURED CITIZEN PARTICIPATION

Neighborhood residents have multiple citizenship: in the neighborhood, municipality, county, state, and the whole nation. In this sense, they are members of several political communities of increasing scale (Hallman, 1977b; Janowitz and Suttles, 1978). These governments have various means for achieving citizen participation beyond the electoral process, and they often look to neighborhoods as collective units for involving citizens in developing public policies and program implementation. This occurs in a variety of ways (Rosener, 1975; Langton, 1978). We'll speak mainly of the city, but the ideas apply to other governmental jurisdictions, too.

Communications is one aspect. It can be one way or two way. The city communicates outwardly by publishing notices and reports and sending out announcements so that citizens, neighborhood organizations, and other groups will know about hearings, official processes, and program actions. Public officials attend neighborhood meetings to report on agency activities. Communications flow the other way as citizens offer their views to public officials by letter, telephone, and personal visits and at public hearings and community meetings attended by these officials. Sometimes these gatherings, especially neighborhood meetings, result in an exchange of views, going beyond one party speaking and the other listening. (For more on hearings, see Checkoway, 1981.)

Advisory committees are a commonplace means for achieving citizen participation. They may be citywide in scope, with membership drawn from many neighborhoods and other interest groups. Or they may function in an individual neighborhood. If the latter, the advisory committee is likely to focus upon a single program area and relate to one particular agency, though some neighborhood advisory committees do consider multiple programs. Members might be appointed by a public official, chosen by designated organizations, selected by citizens participating in a meeting, elected by ballot, or a combination. In the late 1960s when Sherry Arnstein drew a ladder of citizen participation, she placed advisory committees on the lowest run, labeled "manipulation" (1969:218). This particularly reflected experience with urban renewal advisory committees in the preceding decade (Lewis, 1959; Burke, 1966). Since that time, many citizens have learned how to use the advisory role as an entry point for bargaining and exerting influence, and there are public officials who genuinely seek participation and respond to what they hear. Therefore, we should look at the context and the results in determining the effectiveness of advisory committees from the citizen's perspective, not merely the form.

Some neighborhood advisory committees work with public agencies to *develop neighborhood plans*, both physical and programmatic. This tends to follow a conventional planning process: analysis of existing conditions, assessment of need, setting goals and objectives, defining strategies and activities, establishing budgets and timetables (Werth and Bryant, 1979; Urban Systems Research and Engineering, 1980). Ultimately a decision maker other than the advisory committee approves, modifies, or rejects the plan, but neighborhood residents can have an influential role in shaping it. Other planning methods are also used, such as citizen surveys (Webb and Hatry, 1972; ICMA, 1977a; Stipak, 1980), task forces and workshops, charrettes and design-ins, plural planning (each group has its own planner), and mediation (Jordan et al., 1976).

The city budget is the most important policy document adopted from year to year. It determines priorities and allocates resources among programs and neighborhoods. In a few cities, neighborhood organizations have *an official role in budget preparation* (Hallman, 1981b). They determine their needs and offer their recommendations, which are channeled to the budget office and appropriate departments for consideration. They are in touch with the mayor, manager, and city council to push for adoption of their recommendations. (More on this in Chapter 16). In cities without provision for structured participation of neighborhoods in the budgetary process, some neighborhood organizations take advantage of public hearings and use advocacy techniques to impact the budget (League of Women Voters, 1974; Hallman, 1978).

Some public programs make provision for citizen *participation in program implementation*, particularly within neighborhoods. This might relate to selection of personnel, refinement of program details, choice of sites for facilities or beneficiaries of particular activities, and review of program process. The latter might be enlarged to citizen involvement in careful evaluation of program accomplishments, feeding back into the next round of program planning and budget making (Hallman, 1981a).

(Other useful sources on citizen participation include Spiegel, 1968; Cahn and Passett, 1970; Moguloff, 1970; Pateman, 1970; Kasperson and Brietbart, 1974; Falkson, 1974; N. Rosenbaum, 1976; Jordan et al., 1976; P. Marshall, 1977; Lawrence Johnson and Associates, 1978; Langton, 1979; HUD, 1978a; Andersen et al., 1979; ACIR, 1979.)

BARGAINING

From the citizens' perspective, these various practices of structured participation practices have a considerable range of effectiveness in different cities and neighborhoods — from none at all to quite influential. It depends upon attitudes on both sides and upon relative strength in pressing for one's position. Although citizen participation devices don't provide neighborhoods full control and self-governance, they do open opportunities for bargaining. The same is true for interest group activities and political party contacts.

This is important because most public decisions are made through a bargaining process rather than by unilateral fiat of a single individual. Even those who have legal decision-making authority — the mayor, manager, city council, department heads, boards and commissions — take into account the views and influence of others. If you can't be in on the direct bargaining, you want to be represented by somebody who is, or at least to have decision makers aware of your concerns and cognizant that you are a force to be accounted for at the next election, in legal action, or through instigation of protest. In that manner, the neighborhood as an interest group and an electoral constituency is a factor in the broader political community.

Decentralized Administration

In central cities most governmental services provided within neighborhoods are handled by city agencies. In unincorporated suburbs county government plays this role. But also neighborhoods are served by independent school boards, special districts, and state agencies. In many instances, these governmental agencies decentralize their operations to facilities located within neighborhoods, or they divide the city or county into administrative districts, sometimes (though not always) coinciding with neighborhood boundaries or combining several neighborhoods. In this manner the neighborhood serves as an administrative unit, which is another manifestation of its existence as a political community.

CONCEPTS

Let's put this phenomenon into a broader context by distinguishing between noncentralization and decentralization and by defining modes of decentralization. *Noncentralization* refers to an entity which has never been part of a centralized body. Thus, a suburban municipality has this status in relationship to other local governments within the metropolitan area. The same is true for a neighborhood corporation organized by the citizens to initiate a new program or service. *Decentralization* occurs when a function or structure previously centralized is delegated to a unit of smaller scale.

There are three modes of decentralization: political, administrative, and contractual. *Political decentralization* occurs when decision-making authority is assigned to citizens or a representative board of a subarea within the territory of the central authority. *Administrative decentralization* is an internal management process and comes about when subordinates in an organization are delegated discretionary authority to shape program details. *Contractual decentralization* happens when a central agency delegates certain responsibilities via contract to another agency for a stated period of time. It is temporal delegation of specific tasks. (Also see Schmandt, 1972 and 1973.)

PRACTICES

Much of what occurs in the neighborhood as an administrative arena is expressed in the mode of administrative decentralization. Look around any neighborhood and you'll see examples: schools, playgrounds, fire houses, police stations, and perhaps clinics and field offices of a variety of agencies. This happens partly because these facilities by their very nature have to be located around the city and partly for convenience in having personnel located at the point of delivery. Less visible is the division of the city into administrative districts even though personnel may not always be assigned to offices within each district. This is necessary in larger cities in order to achieve manageable units for planning and service delivery purposes. (We'll deal with decentralized service delivery more fully in Chapter 12.)

In most cities, each department draws its own service district boundaries and these rarely coincide with those of other departments. Their criteria are based upon their perception of administrative needs for their kind of service, and more often than not they pay very little attention to neighborhood boundaries as perceived by the residents. A few cities, though, have tried to achieve coterminality for certain key agencies with interrelated programs and have tried to relate this to neighborhoods.

Regardless how departments locate their facilities, deploy their personnel, and draw service district boundaries, neighborhoods find city facilities and services in their midst. Neighborhood organizations and individual residents deal with them by using advocacy techniques, partaking in the agencies' citizen participation practices, and bargaining with their personnel in order to protect neighborhood interests and achieve benefits. Some neighborhoods also seek to obtain policy control over these operations, that is, to bring about political decentralization to go with administrative decentralization.

SCALE

Unit size is an important issue in determining what activities can be administered within neighborhoods, either through administrative decentralization or neighborhood initiative. On this matter, E.F. Schumacher indicated, "For his different purposes man needs many different structures, both small ones and large ones, some exclusive and some comprehensive" (1973:66). For every activity there is an appropriate scale.

Various studies have shown that neighborhood-size units can economically and administratively handle a wide variety of community services, such as refuse collection, police patrol, housing management, street and sidewalk maintenance, employment and training services, public health clinics, elementary schools, playgrounds, social services, and many more (de Torres, 1972; ACIR, 1974; Hallman, 1974 and 1977b). Yet, there are other services which neighborhoods can't manage as well, such as refuse disposal, police laboratory and special investigations, hospitals, sports arena, housing finance, mass transit, and others. But as Schumacher remarked, "What is needed in all these matters is to discriminate, to get things sorted out" (1973:66).

Self-Governance

The fullest expression of a neighborhood as a political community comes when it achieves self-governance. To do this, the neighborhood must have a legal basis for the exercise of power and reliable revenue sources. It requires a policymaking process which is representative of and accountable to the citizenry. It must have an administrative structure capable of carrying out its responsibilities.

ACTUAL

Some neighborhoods within metropolitan areas now have their own governments, taking the form of a general purpose government or a special district.

Suburban units, enclave cities. About one quarter of the metropolitan inhabitants in the United States reside in municipalities, towns, or townships under 25,000 in population. Mostly they are suburban, but some are enclave cities surrounded by the central city. They consist of a single neighborhood, or several neighborhoods usually similar in population make-up. They are established under state law, just as the central city is, and in some states have their own home-rule charters. They have taxing authority and have access to state and federal shared revenue and grants. They have elected governing bodies and either an elected mayor or a manager appointed by council. They are organized into administrative departments with personnel needed to carry out municipal functions. In a very real sense, this is neighborhood government. (See Wood, 1958; Greer, 1962; Perry, 1973; Hallman, 1974; R. Rich, 1980b.)

Special districts. A less complete form of self-governance is the special district, found in some unincorporated suburban areas. Robert B. Hawkins, Jr. has defined special districts as "governmental units deriving their decision-making capabilities from state legislation that are used at the local and regional levels by groups of citizens to solve problems of mutual concern" (1976:9). Although some special districts are adjuncts to city or county government, independent districts are usually governed by their own elected boards and have authority to tax, determine prices for their services, sell bonds, finance capital improvements, and establish their own administrative structures. Fire protection, water supply, and street lighting are some of the functions carried out by neighborhood-level special districts. (Also see Silver, 1981.)

QUASI-GOVERNMENTAL

Although central city neighborhoods lack their own general purpose governments, they have developed instruments which are quasi-governmental in character. One type, the neighborhood council, provides a policy voice in matters ultimately controlled by the city. A second type, the neighborhood corporation, has operational control but functions as a nonprofit corporation rather than governmental body.

Neighborhood councils. In a previous book, I defined neighborhood councils as (Hallman, 1977a:4):

> neighborhood bodies which have some kind of official or quasi-official relationship with local government. They are broad-based organizations of residents from geographic subareas of a city or county. They are usually governed by a representative body, chosen through a democratic process. They focus upon several or many aspects of neighborhood life and not merely on a single problem.

Different cities use different names to describe them, such as community board, advisory neighborhood commission, neighborhood priority board, recognized neighborhood association, and others. They are distinguished from ordinary neighborhood associations by their official or quasi-official recognition by local government as a body to represent the neighborhood.

Some neighborhood councils are created by city charter, others by local ordinance or resolution of city council, some by executive order of the mayor or action by a department. Generally, there seems to be an inverse relationship between (a) the strength of precinct/ward politics and the district election of city council and (b) the creation of officially recognized, elected neighborhood councils. The latter are more likely to emerge where there are at-large elections and where political parties are weak at the precinct level. In contrast, district council members and active precinct workers feel threatened by this alternative method of quasi-political organization and representation and try to prevent its establishment.

For the most part neighborhood councils have only advisory power, but they are built into established processes of local government, such as development of neighborhood plans, capital programming, budget making, and program monitoring. Most cities with neighborhood councils provide them with modest staff support, and some places appropriate funds so that they can hire their own staff or consultant. Some neighborhood councils have initiated self-help activities carried out mostly by volunteers, and a few of them have received grants or contracts to run specific programs during the term of the grant or contract.

Because of their limited authority and the lack of a regular revenue source they can call their own, neighborhood councils don't provide complete self-governance, but they move the neighborhood political community in that direction. (See Chapter 9 for the history of neighborhood councils and Chapter 16 for cases of local operations.)

Neighborhood corporations. During the last 20 years, a sizable number of neighborhoods have organized private nonprofit corporations to carry out particular services or undertake development activities. They go by different names and carry out different functions: community development corporations (emphasizing economic and physical development), neighborhood development organizations (usually with a housing emphasis), community-based organizations (especially for employment and training programs), neighborhood health centers, child care centers, multiservice centers, and others. Most of them got started through federal funding in the period from 1965 to 1980. That remains a major source of their financing (though with cutbacks experienced in the last few years), but they also get support from local and state government, foundation and corporate grants, fees for service, and business enterprises (some of them).

From a local perspective, most neighborhood corporations constitute noncentralization because they have initiated and are carrying out activities which were never centralized. In their funding, they partially enter the mode of contractual decentralization. However, here and there a neighborhood corporation has been delegated operating responsibility for activities previously handled by a local public agency, this being a combination of political and administrative decentralization. We'll examine the work of neighborhood corporations in different fields in Part III.

PROPOSALS FOR NEIGHBORHOOD GOVERNMENT

A number of people have advocated that neighborhoods within central cities should have their own full-fledged local governments. This idea has been around for a long time in the metropolitan reform movement as persons advocating areawide government have recognized that political realities require retaining smaller units as part of a multitiered system; usually this has meant keeping suburban municipalities in place, but occasionally subunits within the central city have been proposed (Jones, 1942; CED, 1970; Bish and Ostrom, 1973; NAPA, 1977; Hallman, 1977b). During the last two decades a number of persons have advocated creation of neighborhood government within central cities out of a concern for local liberty and grassroots control and from a conviction that services can be better administered through small-scale operations (M. Kotler, 1969; Zimmerman, 1972; Hallman, 1974; Morris and Hess, 1975; McClaughery, 1980; Griffin, 1981). At the same time, arguments against neighborhood control have arisen.

Advocates of neighborhood government maintain that it is more economical and efficient to deliver a variety of services by units of small scale. Such administrative units can be more flexible and more responsive to individual needs and particular segments of the population. Providing significant roles for neighborhoods expands the knowledge base for public decision making by involving those affected by problems and services. Neighborhood administration helps build greater self-reliance among the citizenry, and it makes it easier to mobilize neighborhood volunteers, who can help one another. Moreover, participation in neighborhood governance develops citizenship skills which are useful not only within the neighborhood but also in broader governmental arenas. In this respect, neighborhood advocates would substitute "neighborhood" for "municipal" in the following statement of Alexis de Tocqueville (n.d.:book I, 62):

> Local assemblies of citizens constitute the strength of free nations. Municipal institutions are to liberty what primary schools are to science; they bring it within the people's reach, they teach men how to use and how to enjoy it. A nation may establish a system of free government, but without the spirit of municipal institution it cannot have the spirit of liberty.

Those opposed to neighborhood government argue that there is an economy of scale which makes small units too costly and inefficient. They indicate that municipal services require special skills which aren't available in most neighborhoods. Transferring existing personnel to neighborhood units would be costly and not practicable because of existing labor agreements and pension arrangements. Furthermore, it would be awkward to divide off a few activities from an existing service system while keeping centralized other activities which definitely require a large scale of operations. Neighborhoods are parochial and might not be responsive to broader community concerns. This would lead to community divisiveness and fragmentation. Most significantly, neighborhoods have unequal resource bases, so inequities would result if they have to finance their own services.

Advocates counter the last argument by indicating that through shared taxes and grants from wider domains neighborhood inequities could be evened out. An on and on the debate continues. (Most of the arguments were laid out in the late 1960s and early 1970s. For further consideration of pro and con, see National Advisory Council on Economic Opportunity, 1968; Aleshire, 1970; Altshuler, 1970; Shalala, 1971; Schmandt, 1972; and Yates, 1973. For pro, see the advocates of neighborhood government listed above and also Carmichael and Hamilton, 1967; Nisbet, 1975; and Berger and Neuhaus, 1977. For the case against neighborhood control, see Kristol, 1968; Beck, 1969; and Rustin, 1970.)

We'll come back to these issues again in Chapter 16 when we review practical experience with various measures of neighborhood governance and in Chapter 17 when we look to the future. Here we conclude by reiterating that, regardless of how much it controls, a neighborhood is a political community functioning in various ways: as an electoral constituency, an interest group, an administrative unit, and an arena for governance (at least in some neighborhoods).

Exercises

(1) Describe your activities and relationships as voter, member of a political party, and citizen participant in public affairs. Which of these take place within your neighborhood?

(2) What interest groups do you identify with? With which ones do you have an organizational affiliation? Which of these function within your neighborhood?

(3) If you are a member of a neighborhood organization, describe its political activities. If not, find one and describe its political activities. Which of these activities occur only within the neighborhood and which relate to the broader arena?

(4) List all the governmental services, programs, and facilities found in your neighborhood. Which ones have personnel based within the neighborhood? Who directs their work? How does the neighborhood relate to them?

(5) Examine a neighborhood board, council, or advisory committee, preferably one functioning within your own neighborhood, and describe its

powers and basis for authority. On a spectrum ranging from merely advisory to full control, where do you place it?

(6) Think up additional arguments for and against neighborhood government.

CHAPTER 6

A LITTLE ECONOMY

A neighborhood is an economy in miniature, albeit linked to and part of a larger economy. This is so because a neighborhood contains economic resources and is the setting for economic activities. A neighborhood has wealth, enterprises, and a flow of money, goods, and services in and out, and circulation within.

Economic Resources

Look around any neighborhood and you can discover economic resources, though not all of them are immediately visible.

BASIC WEALTH

Considered as a territory, a neighborhood's wealth consists of the value of its physical objects: land, buildings, equipment and machines used to produce goods, and the inventory of goods in process of production. Although neighborhoods differ considerably in total and per capita value, none lacks this kind of wealth.

Buildings and physical facilities. Housing is the foundation of wealth in most neighborhoods, for dwellings are the most numerous structures and in total the most valuable built-wealth. There are, of course, some in-town neighborhoods with numerous commercial and manufacturing structures intermingled. And at least a few commercial buildings are found in most other neighborhoods, though in the suburbs maybe at the edge or concentrated in shopping centers. Industrial buildings aren't as widespread in neighborhoods, nor are warehouses and structures used for specialized services. Indeed, often they are located in nonresidential districts. But some neighborhoods contain structures for light manufacturing.

There are also churches, schools, recreation centers, hospitals, clubhouses, and other buildings owned by government and nonprofit organizations. Also, other physical facilities, such as streets, sidewalks, parks, playgrounds, water mains, sewer lines, gas mains, telephone and cable television lines. They all are part of neighborhood wealth in the broadest sense.

Machines and equipment used in the production of goods and services also belong in the wealth inventory. Economists refer to them and commercial and industrial buildings as capital goods because they provide the physical means for producing consumer goods and services. Money to finance production activities is referred to as capital, or capital investments.

The value of a neighborhood's buildings, facilities, and equipment is determined by the capital invested originally and through the years, extent of deterioration and obsolescence, and by the current market value. For housing, location is an important factor in determining value, with several dimensions. Convenience to places of employment and other attractions and the time and expense of transportation are influential. Social prestige is a factor. Environmental amenities or hazards are considered. For these reasons, identical houses located in different neighborhoods may have different market values.

Land. Another significant part of neighborhood wealth is the land. It was there first and remains a basic commodity. Because land has a finite supply at a particular location, its value changes as demand increases or decreases. Land value is also influenced by permitted uses, such as residential, commercial, or industrial, and the intensity of allowable uses (building height, land coverage, dwellings per acre). Availability of public services — the infrastructure of streets, water supply, sewers, community facilities — affects whether the land can be utilized as desired. A favorable, or unfavorable, social environment influences land values, and also the presence or absence of adverse factors in the physical environment.

Ownership. A neighborhood's wealth, conceived as the value of its physical objects, is fixed in place, but ownership may be near or distant. The owners of houses, commercial and industrial structures, businesses, banks, and manufacturing firms, and vacant land may live in the neighborhood or elsewhere. Neighborhoods can be placed in a spectrum according to what proportion of property is resident- or absentee-owned. Every neighborhood has some resident ownership, and even the most affluent suburb has some of its wealth held elsewhere (for many wealthy people have mortgages on their homes).

Personal wealth. When we turn from a territorial to a personal definition of wealth, we add other factors. Personal wealth consists of the value of (a) land, buildings, and other objects owned, (b) deeds, stocks, bonds, and other debentures representing ownership or an interest in land, capital goods, or enterprises, and (c) money, saving certificates, bank accounts, and the equivalent which can be used to obtain goods and services (such as food stamps and rent vouchers). Residents' collective personal wealth might be located or invested in the neighborhood or elsewhere. This is an important matter in considering the dynamics of the neighborhood economy.

Even poor neighborhoods have savers. Thus, in 1976 residents of East Los Angeles, an eight-square-mile community of 105,000 inhabitants, accumulated $14 million in household savings. Yet, 17 percent of the population were unemployed and 23 percent of the households were below the poverty line (Mahmood and Ghosh, 1979:10).

People. The people living and working in the neighborhood are another significant economic resource. They offer labor, brainpower, technical skill, managerial proficiency, creativity, and entrepreneurship. Even those without much tangible wealth have a contribution to make to the neighborhood economy.

Some residents own, manage, or work for economic enterprises (private, public, nonprofit) operating within the neighborhood. In doing so, they earn their livelihood and help the neighborhood economy to function. Other residents work outside the neighborhood but bring their earnings home. A third group consists of nonresidents holding jobs or running enterprises in the neighborhood but taking their earnings elsewhere. This threeway mixture of work and residency combines differently among neighborhoods. For instance, an ethnic neighborhood with a full range of resident-owned businesses and resident professionals may have a sizable portion of its working population employed within the neighborhood (though usually less than half) while the distant suburban subdivision with hardly any enterprises will have very few residents working there.

There is an interchange of workers among neighborhoods, commercial districts, and industrial areas. For instance, East Los Angeles had 17,150 jobs within its boundaries in 1976. Residents held 7,145 of these jobs and nonresidents 10,005. The nonresidents took away a total of $75 million in wages and salaries, but residents working outside the community brought home $176 million in earnings. This was a net gain of $101 million in wages and salaries (Mahmood and Ghosh, 1979:9).

Among neighborhoods there is considerable variation on the portion of residents employed in and out of the neighborhood, and in what kind of jobs. Some inner-city neighborhoods have quite high unemployment rates, especially for youth and unskilled adults, while in more affluent, outlying neighborhoods virtually everyone who wants to work is employed in well-paying professional, managerial, and technical occupations. This means that some neighborhoods gain more from their human resources than others. This affects how much wealth the neighborhood possesses, the prosperity of neighborhood businesses, the condition of dwellings, and also the social climate of the neighborhood.

Another aspect of a neighborhood's human resources is the many residents who serve as volunteers in neighborhood organizations and institutions, thereby providing useful services. Also, many residents exchange skills and goods without using money. In the larger sense, these are economic activities because they produce goods and services for individual consumption.

ENTERPRISES

The various enterprises functioning within a neighborhood are another important economic resource. They might be privately owned and profit oriented, or nonprofit organizations, cooperatives, or some other alternative form of ownership. Governmental agencies active in the neighborhood can also be viewed as economic enterprises.

Commercial and industrial. Although some neighborhoods are purely residential in character, especially in outlying suburbs, the majority have at least some commercial uses, such as a convenience store, an outlet for a dry cleaner, a dentist, ballet instructor, seamstress, or real estate sales person working at home. Older, inner-city neighborhoods with numerous mixed uses have many more commercial and production enterprises in their midst: shops of many varieties; automobile sales and service; a wide range of other services, such as plumbers, electricians, lawyers, doctors, and morticians; small-scale manufacturing and assembly plants; banks, savings and loan associations, and household credit companies; and perhaps some illegal activities, such as gambling and drug trade.

Some of these enterprises are owned by residents, some not. Ethnic neighborhoods have gone through an evolution of immigrants starting retail businesses and small manufacturing firms and entering the mechanical trades and the professions while retaining residency in the neighborhood. Some who prosper buy homes in other neighborhoods but keep their businesses in the old neighborhood. As the original ethnic population is succeeded by another racial or ethnic group, the newcomers find most of the local businesses absentee-owned. The new residents may take over some of them, or form competitive firms. Further out, many suburban neighborhoods were built more for commuters and weren't intended to provide for many commercial establishments. To the extent that retail businesses are present, they are more likely to be part of a chain and not have resident ownership, though perhaps local management. Production facilities (such as textiles, metal work, electronic assembly, and other light manufacturing) in older neighborhoods are likely to be absentee-owned, and usually not present at all in newer neighborhoods. However, some inner-city neighborhoods have community development corporations involved in commercial and manufacturing enterprises (see Chapter 15).

Commercial and production firms based in particular neighborhoods operate in the context of a wider economy, both in the purchase of materials and equipment and in sales of products and services. They may hire local residents, but they will also draw employees from a wider territory. They are likely to rely upon investment funds derived both within and outside the neighborhood, and profits will similarly go both places.

Public and nonprofit. Other kinds of economic activities are carried out in neighborhoods by governmental agencies and private, nonprofit organizations, though the economic aspect of their operations isn't usually highlighted. Yet, like private enterprises, they have capital investments and use capital goods (buildings and equipment) and labor to produce services of many varieties, and sometimes consumer goods, especially housing. They add to available cash in the neighborhood by hiring residents and sometimes making cash payments to individuals (such as welfare, unemployment compensation) or equivalent nonmonetary awards (food stamps). In some instances, government grants and private donations go to neighborhood-based organizations for production equipment and buildings (capital goods). These various operations and the buildings and land owned by the public and private nonprofit sectors add to the wealth of the neighborhood community.

Governments finance their operations mostly from taxes. Some of this revenue comes from taxes levied directly by the particular jurisdiction, some from grants and revenue shared by governments of wider domain (such as state and federal grants going to cities). Public agencies also sometimes charge user fees and, in the case of public housing, rent. Nonprofit organizations operate on funds obtained from charitable giving campaigns, corporate donations, government grants, investments, and user fees. Except for neighborhood-controlled agencies with their own revenue, most of the funds used by public and nonprofit agencies serving a particular neighborhood are collected from a wider base.

The dynamics of this process differs from the market economy operated by the private business sector. Among other things, the test of profitability is absent, and political factors can intrude in determining which neighborhood benefits most. However, there are similarities because the private, public, and nonprofit sectors all provide consumer goods and services, offer employment opportunities, add to the community wealth, and also distribute ownership of this wealth. A true understanding of the neighborhood economy must consider all three sectors and their interaction.

Economic Activities

So far we have examined neighborhood economic resources mainly in static terms, though here and there we have touched on neighborhood relationships with the broader metropolis. We can learn much more about neighborhood economics by considering the dynamics of economic activities and the flow of resources in, out of, and within neighborhoods.

CASH FLOW

Every household, even the poorest, has a cash inflow. Composite sources include salaries and wages, business profits and professional fees, interest and dividends, sale of goods and services, sale of property, private pensions, welfare and social security payments, gambling winnings and cash return from illegal activities, gifts and donations. Households may also receive free goods and services, such as food stamps, medical care, subsidized housing. Neighborhood businesses and nonprofit organizations also have cash inflow and some of the latter receive donated goods and services. The economic inflow to households, businesses, and organizations comes from sources both within and outside the neighborhood.

Every household also has a cash outflow, going for food, clothing, rent or mortgage payments, utilities, purchase of an automobile and other durable goods, doctors bills and medicine, charitable donations, entertainment, debts, investments, taxes, gambling, and many other things. Neighborhood businesses and nonprofit organizations likewise have a cash outflow. Some of the outflow goes outside the neighborhood, some stays within and recirculates, as illustrated in Figure 6.1.

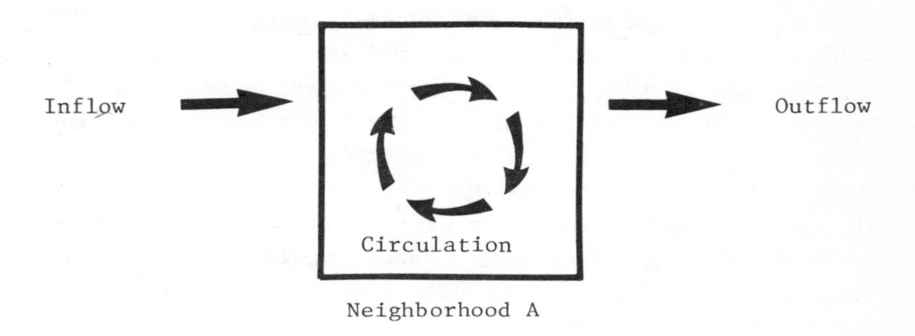

Figure 6.1 Neighborhood Economic Flow

If we traced the economic flow in, out of, and within different neighborhoods, we would find wide variations depending upon the economic strata of the residents, the number and prosperity of neighborhood businesses, ownership patterns, the presence or absence of resident-controlled economic operations, lending policies of local banks, the level of taxation, and other factors. We would discover that even the poorest neighborhood has a considerable cash flow. For instance, 3,000 households with average income of $6,000 per year have a combined annual income of $18 million. How much of this circulates within the neighborhood and how much and how soon it flows outside determines how well the neighborhood is benefiting from its money input. (See Morris and Hess, 1975:69-74; Center for Neighborhood Development, 1980.)

By way of illustration of different patterns of inflow and outflow, we can note a comparison Richard L. Schaffer (1973) made of two community districts of similar size in Brooklyn: Bedford-Stuyvesant, a black community with many poor people, and Borough Park, a predominantly middle-class Jewish district. As Table 6.1 shows, in 1969 residents of Borough park had over twice the personal income than residents of Bedford-Stuyvesant. In both communities more than 80 percent of their earnings come from jobs outside the neighborhood. Bedford-Stuyvesant had somewhat more income from governmental transfer payments, especially public assistance, while the bulk of

Borough Park's transfer payments consisted of social security. Bedford-Stuyvesant received about twice the value of government services, which partially helped to compensate for lower personal income. On the outflow side, Borough Park residents paid more taxes, bought more consumer goods outside the community, and saved more. In Bedford-Stuyvesant gambling and narcotics accounted for a significant cash outflow. When all factors were considered, both communities had a favorable balance of payments, that is, greater inflow than outflow. However, for Bedford-Stuyvesant most of the surplus came from governmental services and transfer payments in excess of taxes, while most of Borough Park's surplus came from the private sector.

Ownership is an important factor. Thus, a study in East Oakland (California) discovered that in 1978 more than half of the rent paid by

TABLE 6.1
Inflow and Outflow of Income in Two New York Community Districts, 1969 (in millions of dollars)

	Bedford-Stuyvesant	*Borough Park*
Inflow of income		
Adjusted gross household income	282.8	615.4
Government transfer payments	121.7	108.2
Consumer credit extension	17.6	28.2
Policy winnings	11.7	—
Government services	166.5	82.7
Business sector	252.4	260.7
Total	852.7	1,095.2
Outflow of income		
Personal taxes	51.5	133.4
Consumption	130.3	248.3
Housing	80.1	91.1
Savings	20.2	65.1
Repayments of consumer credit	16.2	25.9
Policy played, narcotics	47.7	—
Business sector	295.2	346.2
Total	640.3	910.1
Balance of payments (inflow minus outflow)		
Gross surplus	212.4	185.1
Surplus from government (transfer payments, services minus all taxes)	207.2	10.7
Net surplus from private sector	5.2	174.4

SOURCE: Schaffer, 1973:12-16.

tenants went to absentee landlords. Studies in Washington, D.C., revealed that one fast-food store exported over $500,000 a year from its neighborhood, and that chain food stores exported $300,000 per year more than non-chain stores. Also, in the District of Columbia about 87 cents of every dollar spent on energy flows out of the local economy (Institute for Local Self Reliance, 1980; Henze et al., 1979; Batko, Connor, and Taylor, 1975).

For some neighborhoods, the challenge is to increase the circulation of the cash flow within the neighborhood. For instance, in the mid-1970s the Southern Dayton View Neighborhood in Dayton, Ohio had about 9,000 residents with an estimated aggregate income of $36 million, averaging $11,600 per household. A study indicated that if 14 percent of the family budget could be captured by local retail stores, this would generate enough business to support 108 employees (Berry and Bell, 1978; Fleming and Conley, 1977). If residents were hired for these jobs, the neighborhood unemployment rate could be reduced. As we'll see in Chapter 15 when we examine a variety of economic development programs, a number of organizations are pursuing strategies which seek to convert neighborhood cash flow into employment opportunities for the residents.

FLOW OF WEALTH

Some of the economic inflow adds to a neighborhood's wealth, such as through investments in housing. Some of the outflow subtracts from the community wealth, such as closing stores and manufacturing concerns and withdrawing investments. Sometimes the residents' personal wealth flows out of the neighborhood, such as when their savings in a neighborhood savings and loan association finance new developments located elsewhere.

A net inflow of wealth adds to a neighborhood's prosperity while a net outflow causes economic decline. On the latter Stanley J. Hallett has noted (1978:10),

> As structural disinvestment proceeds, different sectors of the economy are affected in different ways. . . .In the private economy, activities tend to decrease. Housing deteriorates, stores become vacant, churches lose members, neighborhood newspapers lose advertising, restaurants and service establishments disappear. As these activities decrease, local jobs disappear with the result that family income decreases. In one neighborhood in Chicago, with a population of 80,000, it was estimated that 300 to 400 jobs were lost in housing maintenance alone when the area became redlined. An additional 500 to 600 jobs were lost as 150 local

> businesses disappeared and a similar number of part-time jobs were lost as other businesses cut back. Local organizations and institutions cut staff by at least 100. . . .This job loss is then translated into loss of family income. The loss of 1000 local jobs translates into a loss of $10 million of neighborhood income, which in turn translates into reduced spending, savings, and tax-paying power.

HOUSING MARKET

The flow of wealth is especially evident in the housing market. Originally the development process draws wealth from elsewhere to pay the capital and labor costs of constructing a new neighborhood. Among the sources are the developer's savings, money from investors, construction and mortgage loans, tax dollars for public improvements, and expenditures of utility companies for electric and telephone lines and gas mains. The new owners draw on their own savings to pay part of their acquisition costs, but they also rely upon mortgages. Their mortgage payments, taxes, and utility payments give the lending institutions, government, and utilities return on their investments.

The sales price of new houses is a combination of land costs of the developer, capital and labor costs of construction, increased land values attributed to quality of construction and desirability of location, cost of financing, and profit for the developer and other investors. Many of these elements derive from factors of supply and demand in the metropolitan economy, but locational matters are rooted in the neighborhood itself.

Over the years housing values change in various neighborhoods, and therefore so does the composite wealth represented by housing. This occurs as an interaction of many factors, such as individual preferences, notions of desirability and undesirability of particular locations, scope of choice of different socioeconomic and racial/ethnic groups, decisions of lenders to invest in or withdraw investments from particular neighborhoods, level and quality of public services, total housing supply in the metropolitan area and in particular locations in relationship to demand, and transportation convenience. As a result, the market value of housing in different neighborhoods will rise, remain stable, and fall in varying patterns.

During the 1970s in prosperous metropolitan areas, for instance, in suburban sections considered desirable by upper income persons, house prices increased ten percent or more each year, and buyers were able to obtain ample mortgage financing. In contrast, in some older neighborhoods with less demand and lack of mortgage financing, housing values declined. Yet, there were neighborhoods where demand was steady but residents and potential incoming buyers discovered that lenders had "redlined" the neighborhood. That is, they either wouldn't make new mortgage loans or offered them on undesirable terms, such as larger down payment, shorter term, or higher interest rate compared to favored neighborhoods. (More on redlining in Chapter 14.)

The dynamics of individual and institutional decisions on housing markets affects the flow of neighborhood wealth. For example, a survey conducted by the Federal Home Loan Bank Board of Chicago in 1972–1973 discovered that many of the city's older neighborhoods got back a much smaller amount of their own savings in mortgage loans than newer suburban communities. Overall suburban residents got ten times the amount of mortgage loan per dollar deposited in savings and loan associations. In the worst case one inner city area had $30 million in deposits in one year but only $360,000 in mortgage loans (Naparstek and Cincotta, 1976:24). Since then federal regulations have reduced such disparity partially but not totally.

In addition to the wealth it provides, housing in many neighborhoods, particularly those with few employers and limited commercial establishments, is the focus of the greatest amount of economic activity. Through construction, sales, rental, furnishings, management, maintenance, rehabilitation, and finance activities, housing provides a sizable number of jobs and a considerable flow of money in, out of, and within the neighborhood.

TRADE

Another aspect of neighborhood economic activities is the part it plays in the broader distribution system which provides raw materials and machinery to producers and delivers goods and services to consumers. Intraneighborhood transactions occur within this framework, usually at the end of the production chain. Thus, a local grocer buys meats, vegetables, canned goods, and other processed foods from wholesalers, who buy from food processors, who buy from farmers. The clothing store, pharmacy, automobile service station, and tavern are similarly at the retain end. But there are also neighborhood businesses engaged in manufacturing, assembling, and processing, such as dressmakers, printers, electronic assembly plants, data processing services, producers of solar heaters. Neighborhood-made products can also be distributed outside the neighborhood. How large a role neighborhoods play in the distribution process varies considerably, depending upon the completeness of the neighborhood economy.

East Los Angeles, referred to earlier, had businesses with an annual sales of $466 million in 1976. Of this amount, $196 million occurred within the community, and $270 million worth of goods was exported. At the same time the community imported goods valued at $252 million. This yielded a net export of $18 million (Mahmood and Ghosh, 1979:9).

As goods and services are distributed, they are exchanged, usually for money but sometimes for other goods and services. Retail sales are one of the most important parts of a neighborhood economy, and retailers are usually the most numerous of neighborhood businesses. Money is the basic medium of exchange, represented by cash, checks, money orders, and charge accounts. The total money income of residents is an important factor in determining prosperity of a neighborhood's economy, and wide disparities exist among neighborhoods in every city and metropolitan area in the United States.

Some people, especially lower income persons, receive and deal with money substitutes, such as food stamps and vouchers for rent and pharmaceutical prescriptions. People who qualify can also receive free or subsidized medical care, school lunches, housing, job-skill training, and certain other services not similarly available to the general public. Sometimes commodities, especially food and clothing, are provided free to poor people.

Neighborhood residents also exchange goods and services through bartering, that is, offering one good or service and getting back another. Much of this is informal as people loan and borrow tools, babysit for one another, trade homegrown vegetables and household furnishings. In these cases, it is one-to-one exchange and duplicates a practice occurring in the earliest, simplest human communities. In other instances, barter groups are formed and use an accounting system which enables a person to provide a service to one neighbor, such as repairing an automobile, and receive an equivalent service, such as plumbing repair, from another neighbor. Although the value of bartering doesn't appear in ordinary economic statistics, it can be an important part of a neighborhood economy (Simon, 1979; Fletcher and Fawcett, 1979; Tobin, 1980; Tobin and Fletcher, 1981).

Organizational Issues

What economic activities a neighborhood can initiate and carry out are determined by a variety of factors, such as willingness, internal resources, availability of external resources, market within and outside the neighborhood for neighborhood products and services, transportation costs, and other linkages with the broader economy. How a neighborhood organizes itself for economic action is an important consideration, and so also is scale, an issue we considered previously in connection with public services.

SCALE

Numerous businesses are small: single proprietors, partnerships, firms with a handful of employees. Repair shops and some other kinds of services can operate at small scale. So can certain kinds of retail establishments, though larger food store chains have the advantage of mass purchasing and marketing. Thus, there are certain types of sales and services which can be handled by businesses located in neighborhoods, serving the residents and owned by residents.

But what of manufacturing enterprises? Barry A. Stein has written (1974:24), "All empirical studies indicated that below a certain size, firms (or plants) are, on the average, technically less efficient, although that 'certain size' is not known with any precision. Moreover, it differs markedly with the specific industry evaluated." At the same time, there are diseconomies of scale from growing too large. In fact, he pointed out (1974:58):

> Large size, in and of itself, is a decided deterrent to work satisfaction (blue- and and white-collar) and motivation. Rates of absenteeism, grievances, and strikes have been shown to be connected directly with size; mental health is inversely correlated. Moreover, competitive efficiency, based on both new product utility and internal motivations in process and technology, is more consistent with small firms in which workers can better understand the relationship of their work to both the organization and its market.

As to whether neighborhood-based plants are likely to be below the threshold of being too small to be efficient, Stein indicated that it depended on the kind of enterprise. However, for a number of industries, a manufacturing firm "employing on the order of a hundred persons may well be competitive. At the very least, one ought not *automatically* to assume that, in any given case, modest size firms will be relatively less efficient" (1974:25). Furthermore, many manufacturing processes (including subcontracting) can be handled economically with considerably fewer employees. (Also see, Morris and Hess, 1975:115-143).

FORM

In keeping with the overall pattern of the American economy, most neighborhood economic enterprises are privately owned. There are also alternatives in which consumers, workers, and residents have a share of ownership, including cooperatives, worker-controlled enterprises, community development corporations, community development credit unions, neighborhood development bank, sweat equity cooperative homesteading, and community investment trusts (N. Kotler, 1978). In addition, there are merchants' associations, local development corporations, and investment organizations which promote and assist neighborhood economic development. Neighborhood land trusts deal with land ownership, and nonprofit housing corporations build, rehabilitate, and manage housing. Various employment and training agencies contribute to human resource development related to economic opportunities, and technical assistance providers help business personnel develop entrepreneurial and management skills.

We'll look at these forms in detail in Chapter 15 when we examine practices of neighborhood economic development. There we'll also deal with the need for capital infusion into poor neighborhoods as a means of bringing about new economic activities. Here we conclude by reaffirming that neighborhoods can indeed by conceived as little economies functioning within the realm of the larger economy which surrounds and encompasses them.

Exercises

(1) List all your personal wealth (see definition, p. 76), or if you don't have much, the wealth of your parents or someone else. Where is it physically located?
(2) Make a listing of your neighborhood's wealth (see separate definition, p. 75).
(3) Chart your personal economic flow, that is (a) the input of money, goods, and free services you received, (b) the goods and services you consume, and (c) the outflow of money and the goods and services you provide others. How many of these transactions occur within your neighborhood and how many occur elsewhere?
(4) Make a similar chart, in general terms, for your neighborhood. Estimate the annual money inflow and outflow and the amount of internal circulation. Give particular attention to methods of recirculation.
(5) List the primary economic activities occurring within your neighborhood. Describe who is involved.

CHAPTER 7

METROPOLITAN PERSPECTIVE

We have seen that neighborhoods have many faces and take many forms. Depending upon their interests and viewpoints, people perceive the neighborhood as a personal arena, a social community, a physical place, a political community, or a little economy. Yet, it is all one neighborhood, merely seen from different perspectives.

WHOLENESS

As a community, the neighborhood is one people even though for analytical purposes we have chosen to consider major characteristics separately. However, neighborhoods vary considerably in their completeness.

All of them are physical places, though some are designed more coherently than others. All are the site of personal interaction, though residents have differing levels of intensity in neighboring activities. All are social communities with networks of relationships and institutional arrangements, but they differ in degree of cohesion. Every neighborhood displays traits of a political community, but some have more self-governance than others. Each one is a little economy, though they vary in their per capita wealth and degree of neighborhood control over economic resources.

These characteristics combine and overlap. Certain physical layouts promote greater social interaction. Social cohesiveness increases the chances of achieving a strong political community. Political strength contributes to gaining greater control over the neighborhood's economy. Adequate economic resources help support neighborhood problem-solving activities.

We'll return to the theme of neighborhood wholeness in Chapter 17. Here we want to look at how neighborhoods within a city and metropolitan area compare and to examine patterns formed by the mosaic. When we look at neighborhoods from this metropolitan perspective, we see two things: differentiation and change. As we further examine what exists, we can define networks and systems tying neighborhoods to the life of the wider metropolis.

Differentiation

No neighborhood is like any other. Each has its own unique combination of social and physical traits. However, some of these traits are similar among certain neighborhoods so that scholars and local planning officials sometimes classify neighborhoods according to key characteristics.

One commonly used classification relies upon three major dimensions: social class, racial/ethnic makeup, and family status/life-style. This derives from social area analysis initiated in California by Eshref Shevsky with Marilyn Williams (1949) and with Wendell Bell (1955). Others have tested and modified the methodology but the threefold scheme remains in use in determining urban residential patterns (Johnston, 1971; Timms, 1971).

We'll look at these three characteristics and then a couple more sets of variables as a means of gaining a better understanding of neighborhood differences.

SOCIAL CLASS

"When societies are complex and service large populations, they always possess some kind of status system which, by its own values, places people in higher or lower position," so wrote sociologist W. Lloyd Warner (1949:8). Contemporary American society defines social status of individuals by a combination of such factors as occupation, amount and source of income, family background, education, community activities, and size, quality, and location of their dwelling. In working out an index of status characteristics for purposes of field surveys, Warner and associates (1949:123) chose four and gave them the following weights: occupation — 4; source of income, such as inherited wealth, professional fees, salaries, or wages — 3; house type, especially size and condition — 3; and dwelling area — 2. The latter was determined by reputation among residents of the locality. Other social scientists have added other factors, such as education, to their measures. The product is a division of the population into social classes. Thus, David L. Birch and associates in their neighborhood studies have projected a fivefold classification of urban American (1979:8): upper class (1.8 percent of the population), upper middle (12.7 percent), middle class (32.0 percent), working class (38.0 percent), and lower class (15.5 percent).

As noted, a person's neighborhood is taken into consideration in defining his or her social class. The reverse can also be done, that is, classifying neighborhoods according to the status of their residents, as determined by their occupations, income level, education, and size and quality of their dwellings. In this manner, the neighborhoods of a city and metropolitan area can be identified according to social classes: upper, upper-middle, middle, lower-middle (or working), and lower. This process is sometimes simplified by using only average income as the determinant because income is associated with many other status factors.

All cities are not alike, however, in their class structure. Therefore, the neighborhood class structure differs from place to place. As Warner noted (1949:23-24),

> Class varies from community to community. The new city is less likely than an old one to have a well-organized class order; this is also true for cities whose growth has been rapid as compared with those which have not been disturbed by huge increases in population from other regions or countries or by the rapid displacement of old industries by new ones. The mill town's status hierarchy is more likely to follow the occupational hierarchy of the mill than the levels of evaluated participation found in market towns or those with diversified industries. Suburbs of large metropolises tend to respond to selective factors which reduce the number of classes to one or a very few. They do not represent or express all the cultural factors which make up the social pattern of an ordinary city.
>
> Yet systematic studies . . . from coast to coast, in cities large and small and of many economic types, indicate that despite the variations and diversity, class levels do exist and that they conform to a particular pattern of organization.

Although cities vary in how residents perceive social class differences among neighborhoods, the people of every American city do recognize neighborhood class distinctions. This affects where they live, would prefer to live, and would never consider living.

RACE AND ETHNICITY

Through a combination of coercion and choice, some racial and ethnic groups concentrate in particular neighborhoods. For blacks, this pattern relates to historic practices of segregation and discrimination, especially as the black urban population has grown in this century. Before that, many northern cities had smaller pockets of black population in different neighborhoods, and southern cities had mixture of blacks (in small houses) and whites (usually in better structures) in the same neighborhood (Drake, 1965). As black urban migration increased, black pockets coalesced into larger ghetto areas which then grew mainly through peripheral expansion. Although there are some stable, racially integrated neighborhoods in a number of cities and suburbs (Bradburn, 1971; Milgram, 1977), in many places a mixed neighborhood is often only a way station to a mostly all black concentration.

Exclusionary practices of realtors, lenders, developers, home sellers, homeowners associations, and governmental agencies have been the contributor to the creation of racially segregated neighborhoods. Other factors are the generally lower income of blacks, thus limiting their housing choices, and a preference among blacks to live among people with whom they feel comfortable and near institutions and commercial establishments catering to their needs. Involuntary and voluntary factors interact to retain black neighborhoods even though legislation has removed many restrictions and economic opportunities have provided more blacks with incomes sufficient to buy homes or rent apartments elsewhere.

Many of the same factors have contributed to concentration of ethnic groups in certain neighborhoods: persons with roots in Mediterranean and East European nations, Hispanics, people from different Asiatic lands and the Middle East, French-Canadians, native Americans. Poverty and limited economic opportunities, especially during the initial period of inmigration, have restricted their housing choices, and social attitudes of others have kept them out of certain neighborhoods. The need and desire to be together, and the practice of the first immigrants attracting and helping later immigrants to settle near them, have led to development of ethnic institutions and distinct subcultures, based in particular neighborhoods. Jews, originally coming from a variety of nations, have similarly coalesced because of their common religious heritage and because of discriminatory exclusion from some residential areas.

Social class and race/ethnicity overlap so that neighborhoods can be designated by combinations: white upper class, black middle class, mixed working class, poor black, white ethnic/lower-middle class, Hispanic with mixed incomes, and so forth. However, as we use these shorthand designations, we should be aware that we are speaking of modal characteristics, that is, are describing what a majority of people in the neighborhood are like, not necessarily everyone. In actuality every neighborhood has some multiplicity of social classes and even more diversity when other variables are taken into consideration, such as occupation, education, religion, and life-style.

FAMILY STATUS AND LIFE-STYLE

Most people change their residence during their lifetime and, when they do, family status and life style affect their housing choice. A major division is between households with children and those without. Families go through different stages — pre-child, child bearing, child rearing, child launching, returned-child, and post-child, and they have different housing needs as they proceed through this cycle (Abu-Lughod and Foley, 1960). Some neighborhoods, especially those with single-family houses and large flats, are particularly geared to the child rearing and launching stages but will also have residents whose children have moved away, and some childless households, too. Other neighborhoods have many small apartments and single people, or adults without children, and sometimes larger households of unrelated adults. Although neighborhoods of these opposite poles — family and singles oriented — can be identified, numerous other neighborhoods have a mixture of child-rearing families and childless households. Thus, this is the least useful of the three dimensions of social area analysis for classifying neighborhoods.

HOUSING AND ECONOMIC CHARACTERISTICS

In addition to social class, race/ethnicity, and family/life-style status, there are other ways to differentiate neighborhoods. City planners and community development officials often classify neighborhood according to condition of housing, relying upon U.S. Census statistics on deterioration, lack of plumbing, and overcrowding or upon their own surveys. In using these data, they translate condition into recommended action strategies, such as clearance, rehabilitation, or conservation, based upon the kind of treatment they think is necessary to improve or preserve housing quality.

Urban economists, having a strong interest in housing markets, often look at property value in classifying neighborhoods. Some of them make use of property assessment data, sales prices of single-family homes, or apartment rent, related to size of units. In developing a model of the urban housing market, the National Bureau of Economic Research has developed a measure of neighborhood quality based upon the quantity of structure services, derived from operating and maintenance expenditures and capital input (Kain and Apgar, 1979:177). Yet, as David Segal has pointed out (1979:8)

> Like some sociologists before them, economists have found household income and race (and after them, life cycle characteristics) to be the

variables that best define neighborhoods or changes in them. What differentiates the work of economists studying neighborhoods from that of sociologists is the explicit role they assign to market processes in transmitting and distributing the consequences of demographic forces and changes in them.

SOCIAL-STRUCTURAL CHARACTERISTICS

Beyond the characteristics of residents, neighborhoods can be described according to patterns of social interaction within the neighborhood and in relation to the wider community. Thus, in studying neighborhoods in the Detroit area, Rachelle and Donald Warren have come up with a sixfold typology of social-structural characteristics, cutting across social class, income, and ethnic lines, as follows (1977:96-97):

> *Integral*: A cosmopolitan as well as a local center. Individuals are in close contact. They share many concerns. They participate in activities of the broader community.
>
> *Parochial*: Has a strong ethnic identity or homogeneous character. Self-contained, independent of larger community. Has ways to screen out what does not conform to its own norms.
>
> *Diffuse*: Often homogeneous setting, ranging from a new subdivision to an inner-city housing project. Has many things in common, but there is no active internal life. Not tied into the larger community. Little involvement with neighbors.
>
> *Stepping-Stone*: An active neighborhood. People participate in neighborhood activities not because they identify with the neighborhood but often to get ahead in a career or some other nonlocal destination.
>
> *Transitory*: A neighborhood where population change has been or is occurring. Often breaks up into little clusters of people; frequently oldtimers and newcomers are separated. Little collective action or organization takes place.
>
> *Anomic*: Really a nonneighborhood. Highly atomized, no cohesion. Great social distance between people. No protective barriers to outside influences making it responsive to some outside change. Lacks the capacity to mobilize for community action from within.

This typology hasn't been tried on any other metropolitan area, so it's not possible to say how universally descriptive it is. Moreover, I'm skeptical of the term "anomic," for it conjures other sociological concepts of "social disorganization" and "normlessness," which I see as outsiders' lack of perception or approval of how a neighborhood people live and function. Nevertheless, the Warrens are right in trying to sort out how various neighborhoods function and in showing that there are differences in key social-structural characteristics.

Change

To further our metropolitan perspective of neighborhoods, we need to understand how they change over the years. The place to start is by exploring patterns of urban growth.

PATTERNS OF GROWTH

In the second quarter of this century scholars postulated three basic models to describe how cities are structured and how they grow: concentric zones, or rings (Burgess, 1925); sectors (Hoyt, 1939); and multiple nuclei (Harris and Ullman, 1945). Since then these models have been tested and debated. Techniques of social area analysis and factorial ecology have offered more refined understanding. "Surprisingly," Larry S. Bourne has written (1982:13), "these three classic models have remained an essential component in describing and explaining urban social patterns in the aggregate, although their interpretation varies widely."

One of the difficulties is that considerable variation occurs among different metropolitan areas so that accurate generalizations are difficult to make. Recognizing this limitation, let us examine how urban areas grow and how this affects neighborhoods. As we do, we'll draw on all three models.

Most new housing is constructed on land never before built upon, and most such land is on the periphery of currently developed areas or beyond. Consequently cities have experienced steady peripheral expansion, growing outward until reaching natural barriers, such as a bay, river, or mountain. This has happened mostly through the actions of many individual builders and companies in numerous tracts and subdivisions, spurred by market opportunities rather than directed by a master plan determining the timing, type, and location of new housing. In addition, public housing authorities have been active since the 1930s and nonprofit organizations in increasing numbers since the 1960s.

Development has often occurred first along major transportation axes: roads, street car lines, commuter railroads, subways elevated railways, and expressways. Then the interstices fill in. The commuter railroad, and to some extent electric street-car lines, made possible new subdivisions separate from the edge of contiguous development. These were consciously designed to form small communities. As metropolitan development later filled in the space around them, they remained as distinct neighborhoods. The use of the automobile as a vehicle for commuting, at first in the 1920s and then exploding massively after World War II, led to numerous subdivisions scattered hither and yon around the countryside. In many places metropolitan growth has spread so far that it has encompassed previously free-standing villages and towns.

Each new tract, subdivision, or project has tended to be fairly homogeneous in housing style and rental or sales price. But in totality over the years new housing has covered a considerable range: for the poor, tenement houses, other low quality housing in earlier years, and public housing more recently; for the working class, modest row houses, tightly crowded frame houses, two- and three-deckers, and trailer courts; medium-size houses at the edge of expansion and in outlying subdivisions for the middle class; larger structures, some of them custom-built, for the upper-middle; and quite large dwellings and mansions for the upper class, almost always customized and often not part of any conventional subdivision. Thus, historically there has been considerable social class differentiation at the time or original construction. However, because of the advent of stringent building codes and land use zoning, the greatest portion of new housing built since World War II has been occupied by higher income groups. Like previous middle and upper income housing, some of it is being handed down to lower income households through a "filtering" or "trickle down" process (Grigsby, 1963; Thompson, 1965; Downs, 1973).

Initially upper-income housing was built close to the center, but as the central business district expanded, rich people moved further out, usually in one direction, though sometimes in a second. This established that section as the "better" part of town and subsequent development of upper class housing has occurred in that sector. In the days before the automobile, working class housing was nearly always built near factories and other places of employment, ordinarily near the center of town or along railroad axes. Middle class housing was constructed along road and streetcar lines and filled the space between. Upper-income suburbs on commuter railroads sprang up beyond the area of peripheral expansion, and previously separate town became nuclei for new development, as did some new, outlying commercial centers and industrial plants. Over the years as upwardly mobile families bought new housing in outlying sections, the older neighborhoods they left were occupied by lower-income groups. Meanwhile, in and near the central business district older housing was torn down for commercial expansion, and in the past 35 years in many cities for high-rent apartment buildings. And during the last two decades in quite a number of cities, higher-income households have taken over and refurbished older housing near the center.

So now, land use in every American metropolis is differentiated into commercial, industrial, institutional, and residential areas, with some intermixture. Generally the most intense use is found in the central business district. Residential densities tend to be highest in the surrounding ring and get lower as the metropolis spreads outward. But here and there now-absorbed older town and other commercial and institutional nuclei break into the concentric pattern. Upper class preference for certain areas and traditional locations of working class housing introduce sectorial variations. So therefore, a general concentric pattern prevails but is modified by natural barriers, sectorial differentiation, and subcenters of growth. Overlaying all this is the concentration of racial and ethnic minorities in certain sectors. All these factors influence the way neighborhoods change.

STAGES OF CHANGE

Through a period of years neighborhoods face the prospect of change as the housing gets older and original occupants die or more away. In many situations structures are well cared for and new occupants are similar to the previous ones so that the physical conditions and social class characteristics of the neighborhood remains the same for a long time. But a neighborhood can experience significant population shifts, changes in housing quality (deterioration or improvement), and modification of land use patterns. This has led Anthony Downs to observe (1981:3):

> A life cycle is evident in many neighborhoods. They evolve from births as new subdivisions occupied by relatively affluent households, through middle age, when they shelter relatively less-affluent households but remain in good condition, to their deaths through decay and abandonment by the poor households that finally occupy them. This phenomenon is related to the concentration of poverty.

In studying this life cycle, Downs has defined a fivefold typology of stages of neighborhood change. He ties it to neighborhood decline, defined as (1981:61) "increasing physical deterioration, reduced social status, greater incidence of social pathologies such as crime, and a loss of confidence among investors and property owners in the area's future economic viability." The five stages are (1) stable and viable, (2) minor decline, (3) clear decline, (4) heavily deteriorated, and (5) unhealthy and nonviable. Not every city has neighborhoods in the third and fourth stages, especially the newer cities in the Southwest and West and medium-size cities in the Midwest and East. Furthermore, according to Downs, neighborhoods can change in either direction along the continuum, improving as well as declining.

During the 1970s there was considerable debate on causal factors of neighborhood decline. Public Affairs Counseling (1975), a group associated with Downs, indicated that the key factor was decisions of individual households whether to stay or move and, if the latter, where they might move to. However, Downs' associates also recognized that decisions of bankers, brokers, and governmental officials had an influence. In contrast, Arthur J. Naparstek and Gale Cincotta (1976) assigned responsibility for decline to decisions of lending institutions to disinvest in neighborhoods with an ethnic or black majority, older housing, and lower income occupants. A more balanced judgment, offered by Roger Ahlbrandt and James Cunningham (1979:26) and by Downs in his recent work (1981:62), is that multiple actors are involved (for a list, see Chapter 14). They interact in different ways in different neighborhoods and at different times in the local and national economy. Social values, especially racial and class prejudice, of persons in financial institutions and governmental agencies and of present occupants and households who are considering the neighborhood, can be significant.

Market perception is another factor, and it interacts with housing conditions as Rolf Goetze has pointed out (1979). Housing in a neighborhood can be in good, fair, or poor condition. Separately people can perceive the housing market there as rising, stable, declining, and rapidly declining. A rapidly declining neighborhood with poor housing is ripe for abandonment and arson while a rising neighborhood with housing in only fair condition is a good candidate for reinvestment and gentrification. This is shown in Figure 7.1.

We'll return to this subject in Chapter 14 when we examine methods used to achieve neighborhood preservation and revitalization.

RACIAL CHANGE

Over the years as people move out of and into particular neighborhoods, those who leave may be succeeded by similar or dissimilar successors. Where the latter is predominant, the neighborhood's population changes in such attributes as income level, occupation, ethnicity, and race. In American cities of the past 35 years black

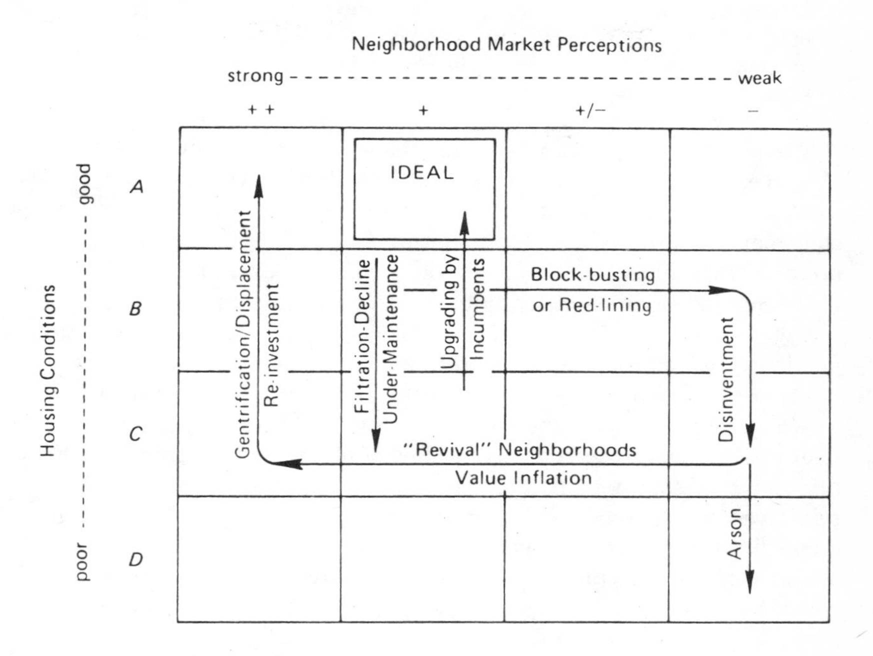

Reprinted with permission from Rolf Goetze's *Understanding Neighborhood Change: The Role of Expectations in Urban Revitalization*, Copyright 1979, Ballinger Publishing Company.

Figure 7.1 Interaction of Housing Conditions and Neighborhood Market Perceptions

succession in previously white neighborhoods (and in some localities Hispanic succession to Anglos) has been a conspicuous trend.

As explained by Howard Aldrich (1975:342):

> The precondition for succession is set when the established residential group in an area no longer replaces itself, whether because of upward mobility into "better" areas, stage in the life cycle, or any number of other conditions. The withdrawal of an established group opens up opportunities for expansion into the area by a new group seeking new housing because of upward mobility, crowding in their previous area, or perhaps the conversion of their old area into new use, as in urban renewal. Real estate agents and brokers are often facilitators and mediators in this process, acting as gatekeepers for the racial

> homogeneity or heterogeneity of a neighborhood. . . .Once the process of succession is begun it is rarely if ever reversed, and succession follows a continuous line of expansion is the established group recedes and the expanding group follows after them.

The market mechanism of racial change has been called arbitrage (Leven et al., 1976; also see Downs, 1981:86-102), a term derived from a stock market practice of simultaneous purchase and sale of the same or equivalent securities in order to profit from price discrepancies. In housing this reflects a dual market, white and minority (black, Hispanic), wherein minority households pay more for housing in a transition zone than whites. The transition zone moves further into previously all-white neighborhoods as higher-income minority households are willing and able to pay the higher price, or as investors buy dwellings once occupied by whites for rental to minority persons. This dual market rests upon restricted access of blacks and Hispanics to the full range of housing opportunities throughout the metropolis. It is reinforced by the tendency of most people, black and white alike, to prefer living with people who are similar, and in this regard, racial and ethnic similarity, income level, and other status factors are important considerations.

Linkages

Neighborhood change occurs because of social and economic forces operating within the metropolis and the nation. This clearly shows that no neighborhood is an island, completely isolated from its broader environment. Instead neighborhoods and their residents are linked to other neighborhoods and to the metropolitan community. This occurs through physical linkages and people's perception of city form. It also comes about through social networks and functional systems.

CITY FORM

Our visualization of city form enables us to relate our neighborhood to the wider city and metropolis. Kevin Lynch has defined five elements of city image which facilitate this process (1960:47-48):

> *Paths*: channels along which the observer customarily, occasionally, or potentially moves.
>
> *Edges*: linear elements not used or considered as paths by the observer, such as shores, railroad cuts, edges of development, walls.

Districts: medium-to-large sections of the city, recognizable as having some identifying characteristics.

Nodes: strategic spots in a city which an observer can enter, such as junctions of paths or concentrations of some characteristics.

Landmarks: a reference point considered to be external to the observer, such as a building, sign, store, or mountain.

We perceive our own neighborhood as a distinct district and recognize other neighborhoods as different places. As we cross over the edge of our neighborhood, our paths lead us toward particular nodes, guided by recognizable landmarks. We probably don't know every inch of territory of the whole metropolis, but in our mind's eye we have some notion of the wider urban form and how our neighborhood fits in. However, as Lynch noted by comparing Boston, Jersey City, and Los Angeles, cities differ considerably in how clearly their physical layout facilitates residents' capacity to comprehend the shape and texture of the metropolis. (On cognitive mapping, see Davis and Stea, 1977: on children's perceptions, see Mauer and Baxter, 1972).

SOCIAL NETWORKS

In Chapter 3 we considered how social networks function within neighborhoods. They don't stop there, for we have social ties with many people living outside our neighborhood: relatives, friends from childhood and school, people we work with, persons with whom we share interests, members of clubs, churches, synagogues, political parties, and other associations. Friends may be dispersed in various parts of the metropolis and beyond (Wellman and Leighton, 1979). We keep in touch with them by telephone and mail, at work or meetings, by doing things together. Upper classes tend to have more extensive "nonspatial" relationships than lower classes, who are more neighborhood bound. Other variables are stage of life, place of work (or not working), and extent of racial or ethnic homogeneity and solidarity in one's neighborhood. As just about all urban residents engage in neighboring and have neighborhood friends, so also they have social ties beyond the neighborhood. We are residents of both our neighborhood and the metropolis. Our social networks provide connections.

This carries over to where we turn for help in solving personal problems. As noted in Chapter 3, Donald I. Warren, in studying helping networks, has traced linkages people have with their immediate family, kin, friends, neighbors, coworkers, voluntary associations, some located within one's neighborhood, some outside. He found that the neighborhood as a geographical location wasn't the center of resources in problem-solving. "Rather, it tends to function as an arena in which an individual can gain access to various kinds of resources through the social network of other neighbors (1981:184)." Having this interest, he has avoided defining neighborhood as a limited territory but instead spoke of the "neighborhood context":

> A neighborhood context refers to the social organization of a population residing in a geographically proximate locale. This includes not only social bonds between members of the designated population but all bonds that group has to non-neighbors as well (1981:62).

SYSTEMS

As we can identify personal linkages between people in neighborhoods with others in the broader metropolis, so also we can observe that functional activities occurring in neighborhoods are parts of larger functional systems.

A system is a group of things which regularly interact so as to form a unified whole. These things may be organs of the body, parts of a machine, specific community services, or ideas of the mind. Whatever the elements, they form a system if they are related to one another. They are interdependent. They function together. They constitute an integrated whole, or at least have that potential.

We can trace such systems for virtually every function of life and can identify where neighborhoods fit. There is the food chain of production, processing, wholesaling, retailing, and finally consumption at homes in neighborhoods. Running water in homes comes from a public water system. A sewerage system and a refuse collection-and-disposal system carry away waste products. A transportation system consists of alleys and neighborhood streets connected to thoroughfares and expressways, and it makes use of private automobiles, taxis, buses, and rapid transit vehicles. Housing production, marketing, and finance are based upon a complex set of system relationships.

Neighborhood services are part of larger service systems. After completing the neighborhood elementary school, pupils go to a middle or junior high school and then to a high school, tied together by a community-wide school system. Neighborhood-based health services, such as those provided by private physicians, clinics, visiting nurses, and sanitation inspectors, relate to hospitals, public health laboratories, and health insurance companies serving a much larger area. Police patrols in the neighborhood have command communications and ties with the city police department, and even in a neighborhood-size suburban municipality with its own police force, the police relate to a wider criminal justice system consisting of prosecutors, courts, and prisons. We could go through other service systems and find similar relationships.

The same idea can apply to the political community and governmental organization. In another book (Hallman, 1977), I have referred to this as "local federalism." This concept draws upon James Madison's observation that our national and state governments are "but different agents and trustees of the people, constituted with different powers, and designated for different purposes" (n.d.:304-305). Thus, neighborhood political instruments, whether they be advocacy organizations, advisory councils, or full-fledged neighborhood governments, serve a distinct purpose in a governmental system which also has municipal, state, and national governments serving other purposes.

The little economy operating within the neighborhood is part and parcel of the metropolitan, regional, and national economies. Cash and wealth flow in and out of neighborhoods, as well as circulate within. Neighborhood sale of goods and services and housing market transactions are strongly affected by the wider economy.

Neighborhoods are small communities within a larger community. Linkages come through social networks and systems. The systems themselves often relate to one another through an even more encompassing network of systems. This is the metropolitan context in which neighborhoods exist.

Exercises

(1) Describe your neighborhood in terms of social class, race and ethnicity, family status and life-style. How much homogeneity is there? What range of traits?

(2) Make a similar description of two or three neighborhoods.

(3) Describe the changes which have occurred in your neighborhood since 1920 (or if newer, since it was developed). Pay attention to population makeup, physical conditions, political and economic activities. How do these changes relate to models outlined in this chapter?

(4) Look again at your social networks, charted in the exercise for Chapter 3, and examine the citywide and metropolitan connections.

(5) Describe several major systems functioning in your neighborhood, what they do there, and how they are connected to components in a wider geographic area.

(6) Rate the importance of neighborhood to you in the main categories we have considered.

	Not Important	Moderately Important	Very Important
Personal arena			
Social community			
Physical place			
Political community			
Little economy			

(7) Rate your neighborhood in a variety of traits such as:

Social community	Fragmented	Cohesive
Physical place:	Vague identity	Clearly defined
Political community:	Powerless	Powerful
Little economy:	Weak	Strong
How these factors interrelate	Disjointed	Integrated

(8) Estimate how different kinds of persons would rate the importance of their neighborhood (as in exercise 6).

(9) Rate other neighborhoods (as in exercise 7).

(10) What factors contribute to neighborhood differences?

PART II

ORGANIZING FOR NEIGHBORHOOD ACTION

CHAPTER 8

APPROACHES BEFORE 1960

In Part I, we have seen that neighborhoods are a natural phenomenon. As social beings, we create our own personal neighborhoods, usually small in size. Through the fibre of social networks and our participation in activities of neighborhood organizations and institutions, we are connected to one another in neighborhoods functioning as social communities. Neighborhoods are also physical places, political communities, and little economies. Neighborhoods are organic because they develop naturally from our lives as residents of particular territories.

We have also observed that neighborhoods exist in a metropolitan setting, tied together by social and institutional networks and served by systems functioning in a broader territory. People circulate throughout the city and metropolis for many of life's activities. At the same time their neighborhoods are impacted by decisions made elsewhere and are influenced by societal forces.

With this conceptual framework established, we are now ready to turn to the practical. For this purpose, in the next two chapters we'll look at the history of how residents and concerned outsiders have organized to deal with neighborhood problems. Then in Chapter 10, we'll draw upon this experience to summarize strategic choices neighborhoods have for moving into action, and we'll review the principal techniques of community organizing.

Use of Associations Before 1860

"Americans of all ages, all conditions, and all dispositions form associations." So observed Frenchman Alexis de Tocqueville as he reported on his travels in the United States in the 1830s (n.d.: book 2, 114). This phenomenon had occurred for at least a century before his visit, and in the ensuing years, the use of associations has multiplied endlessly.

By association, we are speaking of an organization established to achieve shared objectives of its members. It is voluntary in nature, not governmental. It is not an economic enterprise, though an association might seek to influence economic policies. In our discussion we also exclude political parties because of their direct connection with the electoral process tied to government.

Many functions we now perceive as predominantly governmental used to be the province mainly of voluntary associations. For instance, in Philadelphia in 1739 Benjamin Franklin organized the first voluntary fire brigade in the United States, and it wasn't until 1837 that the first paid, public fire department was set up, in Boston (Lane, 1967). For police protection the colonies and the original states used the European watch system, requiring male citizens to stand night watch without pay, though some wealthy men hired substitutes and around 1800 Boston began paying 50 cents a night. In 1833 Philadelphia became the first city to establish a daytime force, followed by Boston in 1838, and New York a few years later (H. Locke, 1977).

When cities were small, voluntary fire companies and town watches weren't consciously organized along neighborhood lines. But here and there associations formed with a territorial orientation. For example, in 1844 owners of 28 homesites surrounding Louisburg Square on Boston's Beacon Hill established the Committee of the Proprietors of Louisburg Square for maintenance of the park area (Urban Land Institute, 1964:39). In the same decade Jewish communities in Eastern cities began forming literary and educational associations for young men, and in 1854 the first Jewish community center opened (Millman, 1960:92). In this period other immigrant groups formed their own ethnically oriented associations. Although not territorially restricted, they functioned as neighborhood institutions because of the geographic concentration of the particular population.

1860 To 1918

During the last half of the 19th century, neighborhoods began to become a more conscious focus of individuals and organizations concerned about improving living conditions. Some initiators came from the better-off segment of society who, for religious and humanitarian reasons, wanted to help the underprivileged. Other initiators lived in the poor neighborhoods. Their response took several forms, which we sample.

CHARITY ORGANIZATIONS

Buffalo, 1877. Each year more and more people are pouring into Buffalo. Many are coming from Europe, passing through New York City and heading for the western end of the state. Others are rural Americans.

Population was 117,000 in 1870, and some are predicting it will go beyond 150,000 in 1880. Poverty prevails among most of these new residents, so Buffalo, like most other Eastern cities, has a private charity organization to help relieve the distress. Now, with the city spreading out, the charity organization has decided to divide its operations into districts corresponding to police precincts. This will get charitable services closer to the people who need them.

Establishment of charity organizations was a response to the misery and despair which accompanied industrialization of the United States in the years following the Civil War. In part, they were relief agencies, but many of the founders were also concerned with what they perceived as "pauperism," that is, basic character which seemed to keep some people poor. One remedy was the use of friendly visitors and other means of promoting neighborly intercourse. But as the charity movement evolved, it paid increased attention to underlying causes of poverty and resulting social conditions — crowded and insanitary housing, tuberculosis, other health problems, child labor, sweat shops, and other industrial ills (Dillick, 1953:28-32; see also Watson, 1922).

With their concern for the poor, it was natural for charity organizations to focus their services in neighborhoods where poor people lived. Buffalo was the first to set up district units, but other cities soon followed. In Philadelphia in 1878, 23 charity societies were organized in wards or groups of contiguous wards. Each society raised and disbursed its own relief funds and elected two delegates to a central board. Boston had district committees of residents and also district conferences involving representatives from private and public agencies working among the poor (Dillick: 32-34).

By the early years of this century, the efforts initiated by charity organizations were evolving into citywide councils of social agencies. This produced greater centralization of operations and more emphasis upon professionalism. District committees remained, but they tended to be dominated by agency staff and professional people living in the neighborhood. Some interest in neighborhood organizing remained but increasingly it came from other sources.

SETTLEMENTS

New York, 1886. As the primary port of entry for European immigrants, New York City has developed an extensive network of charitable services. Within the past year some concerned citizens have heard about a new approach in London where in 1884 a parish vicar invited a number of university students to settle in Toynbee Hall. There they are sharing life in a deprived area in order to gain better understanding of working

> class conditions and to enlist the more fortunate in altering those conditions. Now the first American adaptation of the settlement house is about to open: Neighborhood Guild on Eldridge Street on the Lower East Side.

This idea caught on quickly and spread to many other American cities. Hull House in Chicago opened its doors in 1889, and by 1900 there were 103 settlements in the United States. According to Sidney Dillick (34-35):

> The settlement began as a center established by advantaged persons who desired to help the people of a deprived neighborhood, and developed rapidly into a significant social institution. . . .The settlement assumed a special responsibility for all families living within the radius of a few blocks of the settlement house. It also sustained a general relation to the larger district encircling about the neighborhood. It was concerned with developing institutional resources suited to the needs of a working-class community. This included relief of distress, removal of unsanitary conditions, care of neglected children, and recreation.

Settlements were among the first to sponsor such activities as well-baby clinics, playgrounds, kindergartens, day-care for children of working mothers, public health nursing, mental health clinics, and various forms of adult education (Hillman, 1960:iv). And as settlement workers got to know their neighborhoods and the needs of residents, many of them were drawn into social reform. Some sought to mobilize neighborhood forces, and a few tried to help residents develop self-directed organizations. In Boston, for example, settlements helped organize 16 district improvement societies, which chose delegates to the citywide United Improvement Associations. Settlements formed their own federations. And in a variation, Los Angeles set up six municipal settlements, actually a residence on a playground with club activities, dances, a branch library, and home nursing services (Dillick: 38-54; also see A. Davis, 1967).

SCHOOL COMMUNITY CENTERS

> *Rochester, 1909.* It's been two years since the Rochester Board of Education appropriated funds to use 16 school buildings for civic and social purposes. They have been serving both youth and adults. It worked so well that last year a citywide federation of school-based civic clubs was formed. Unfortunately that was their downfall. Ward politicians are fearful of the competition, so they have cut off funds for the school centers. Now this experiment is coming to an end.

The Rochester experience, though, attracted wide attention. It seemed so natural to use these buildings, paid by public funds, for broader public use. Many other cities took up the cause. By 1911, 48 cities were using 248 school buildings in this manner, and by 1921 the number of cities had grown to 170 (Dillick: 58-66; also see Gluek, 1927; Fisher, 1981).

Initially school centers were mostly paternalistic operations, but after a while some of them developed self-governing committees, consisting of a representative from each member club. In New York through the influence of the People's Institute, school centers were used as a base for forming neighborhood organizations. As Dillick explained this philosophy (1953:61):

> A community center was not defined as a building or as a set of activities, but rather as an organizing center for the life of the neighborhood. The public school was regarded as the natural place for the community center for practical reasons, since school buildings were found in all neighborhoods and were used only half the time. There were equally important reasons of principle, since the school belonged to the public and was the most important agency of the state for spreading knowledge and fostering civic ideals. The community center worker was regarded as a neighborhood leader; he was on the job continuously; he stimulated the community to develop its own activities; and he showed how they could pay their way. The successful community center, it was believed, required full-time leaders or their equivalent.

The school center movement peaked in the mid-1920s. By then they were becoming mostly service organizations. Other community buildings and church social centers came into use and lessened the demand for school facilities. Moreover, the ideal of self-governing school centers never gained widespread application, and other methods developed to organize city neighborhoods, as we'll see momentarily.

INDIGENOUS ORGANIZATIONS AND INSTITUTIONS

During the same period immigrant neighborhoods had other kinds of institutions and organizations. Roman Catholic churches served specific territories (parishes) and particular nationality groups, and Lutheran and certain other Protestant churches also had a nationality orientation. Jewish community centers and other ethnic organizations related to residents concentrated in particular neighborhoods. Black churches were the most important institution for their communities, though there were also lodges and organizations focused on civil rights and black nationalism. Ward and precinct committees fulfilled social service functions. Most of these indigenous organizations didn't articulate a neighborhood emphasis, but because of population patterns they functioned as neighborhood-based institutions.

HOMES AND CIVIC ASSOCIATIONS

Mission Hills, Kansas, August 20, 1914. Five owners of new homes in this budding subdivision just across the state line from Kansas City, Missouri have received a charter for the Mission Hills Home Company. They did this to fulfill a requirement of a declaration they signed when they purchased their homes from the J. C. Nichols Company. The homes association will assess them a proportionate share of the costs of maintaining public spaces, trash pick-up, and recreational facilities. Their deed also contains restrictions on the use of the property, such as permitting only single family occupancy, regulating architectural style, and prohibiting construction of out buildings.

In the use of restrictive covenants and promotion of homes associations, J. C. Nichols, developer of Kansas City's Country Club District, borrowed and improved techniques developed during the 1890s in Roland Park, just outside Baltimore (annexed in 1918). The developer, Edward H. Bouton, and his site planner, George E. Kessler, had come from Kansas city, and when Nichols initiated his first development in 1905, he studied the Roland Park experience. There the Roland Park Civic League formed in 1898 and became the principal stockholder of the Roland Park Roads and Maintenance Corporation to provide the public maintenance services Baltimore County wasn't equipped to handle. After several years of experience, Nichols decided to build this approach into the basic scheme of all his developments (Urban Land Institute, 1964:43-46, 51-54).

A few other new subdivisions of this period also had homes associations to maintain public space, and many more used restrictive covenants to exclude such groups as Jews, Negroes, and Orientals (these prevailed until 1948 when the U.S. Supreme Court ruled that they are unenforceable in courts of law). Where the city or county took care of streets, refuse collection, and parks, residents set up civic associations, usually organized to coincide with subdivision boundaries, and in the second decade of this century, numerous block associations formed in New York City. By and large these were middle and upper-middle class organizations, functioning as interest groups to protect property values, keep up appearance, ward off detrimental influences, and maintain social homogeniety, but some of them sponsored recreational activities for children and youth and undertook other community projects.

1919 To 1944

By the 1920s community councils of professional workers and leading citizens were evolving into centralized councils of social agencies. But many continued to function in neighborhoods and multi-neighborhood community districts. They weren't resident controlled, but they did represent decentralization of social service coordination. In California they took another course as they focused on juvenile delinquency.

AREA COORDINATING COUNCILS

> *Berkeley, California, 1919.* For several months Chief of Police August Vollmer and assistant School Superintendent Virgil Dickson have had lunch together once a week to discuss problems of children known to both of them. Now they have invited other officials and social workers to join them so that they can develop and carry out plans of adjustment for individual children. They have decided to refer to their group as the "coordinating council."

Sixteen years later the idea caught on in Los Angeles County, where by the mid-1930s 60 area coordinating councils were operating to deal with juvenile delinquency. The same approach spread to other cities in California and around the United States. By 1936 there were over 250 local coordinating councils in 163 cities in 20 states, and they got together in a national conference (Beam, 1935, 1936).

As developed in Los Angeles County, area coordinating councils divided into three committees. An adjustment committee brought together representatives of the police, probation department, schools, health and welfare department, and case work agencies. A character building committee consisted of persons from the playground department, libraries, churches, boy and girl scouts, YWCA, and YMCA. An environment committee had members representing civic organizations interested in child welfare, such as PTAs, women's clubs, men's service clubs, American Legion and its auxiliary. Each coordinating council had a small staff and typically served a secondary school area, thus providing a geographical basis for bringing together public officials, social service agency personnel, and representatives of citizen groups.

CHICAGO AREA PROJECT

> *Chicago, 1933.* Help wanted. The Illinois Institute for Juvenile Research is seeking residents from certain Chicago neighborhoods to staff a new program aimed at reducing juvenile delinquency. Applicants must live in the neighborhoods served by the program. They must have thorough knowledge of the people and organizations in their neighborhood and a natural capacity for working with adults and youth. Duties include organizing civic committees to deal with delinquency and fostering youth welfare activities. The jobs have no minimum education requirements but workers will receive training and supervision from staff professionals of the Institute.

In this manner Clifford R. Shaw began the Chicago Area Project. It was a product of many years of sociological research into the causes and nature of juvenile delinquency in which Shaw and his associates carefully documented historic patterns of delinquency and found no correlation with nationality or race, as some claimed. Rather in that period the neighborhoods with highest rates of delinquency were populated predominantly with recent migrants from rural areas of the Old World. Their efforts to adapt their social institutions to the urban setting were only partially successful. As a consequence, youth subcultures emerged with their own variant traditions, often characterized by criminal behavior. Nevertheless, Shaw maintained, delinquency often represented an attempt to achieve status as a human being, no matter how antisocial it may appear. Thus, it was an adaptive behavior in a setting where the machinery for social control had broken down.

The Chicago Area Project had two other sociological postulates, according to Solomon Kobrin, who joined the Chicago Area Project when it was eight years old. First, "the source of control of conduct for the person lies in his natural social world." It follows that delinquency prevention activities must gain the sponsorship and support of neighborhood adults who are part of the local social order. Second, "people support and participate only in those enterprises in which they have a meaningful role." Even in areas with high delinquency and seemingly immense disorder there exists "a core of organized communal life centering mainly in religious, economic, and political activities." Here was a strength to build on (Kobrin, 1959:22-23).

Based upon these assumptions, the Chicago Area Project stimulated the organization of neighborhood groups to lead the way with youth activities. To accomplish this the Illinois Institute for Juvenile Research, the overall sponsor, set about hiring indigenous workers as the key actors to organize the adults and reach out to delinquent youth.

They looked for residents who possessed a natural knowledge of the local society, who could speak the language of the people, and who would have easy access to the youth. Staff sociologists trained and supervised them, but the workers had considerable latitude within the neighborhood.

So also did the local organizations, for the Chicago Area Project encouraged independence for the neighborhood groups. These groups could nominate residents to become their staff, or veto staff proposed by the Area Project. Neighborhood staff were expected to identify with the local group rather than the Institute. And the Area Project respected decisions of neighborhood groups, even if they appeared unsound. Thus, hiring of indigenous workers and fostering independence of neighborhood groups were hallmarks of the Chicago Area Project.

The neighborhood groups sponsored a variety of activities. Recreation programs were a common fare, not too different in content from those offered by traditional social welfare agencies but noted for considerable use of neighborhood volunteers and improvised locations in storefronts and unused spaces in churches and police stations. They sponsored community improvement projects related to schools, sanitation, traffic, law enforcement, and housing conservation. And they carried out activities directed specifically to delinquent children, youth gangs, and offenders returning to the neighborhood from penal institutions.

THE ALINSKY APPROACH

> *Chicago, July 14, 1939.* People are starting to arrive for the first meeting of the Back-of-the-Yards Neighborhood Council. It's a tough neighborhood next to the stockyards, where many of the people work. Organizer of the meeting is Saul D. Alinsky, originally assigned to the neighborhood by Clifford Shaw as an outgrowth of the Area Project. A criminologist by graduate education, Alinsky worked for a while at the Joliet penitentiary then became a labor organizer for John L. Lewis who has put together the new Congress of Industrial Unions (CIO) after some raucous battles. From this varied experience, Alinsky has developed some strong ideas about the need for people's organizations to fight for justice and democracy. Now after months of preparation, the first one is ready for launching as the indigenous leaders in Back-of-the-Yards are calling the meeting to order.

Alinsky's careful organizing paid off, and the Back-of-the-Yards Neighborhood Council grew into a vigorous organization. Next year he set up the Industrial Areas Foundation and began sending organizers to other working class neighborhoods in northern industrial cities and to

Mexican-American communities in the Southwest. They weren't always welcomed by established community leaders. Sometimes they were arrested or thrown out of town. Alinsky himself spent some time in jail as a consequence of vigorous organizing.

Alinsky rejected the approach of the conventional community councils, which "have largely confined themselves to coordinating professional, formal agencies which are first superimposed upon the community and subsequently never play more than a superficial role in the life of the community" (1969:65). He felt they didn't involve the true indigenous leaders and that their style was too tame for dealing with problems of poor communities. In contrast, he wrote:

> A People's Organization is a conflict group. This must be openly and fully recognized. Its sole reason for coming into being is to wage war against all evils which cause suffering and unhappiness. A People's Organization is the banding together of large numbers of men and women to fight for those rights which insure a decent way of life. . . .
>
> The building of a People's Organization is the building of a new power group. The creation of any new power group automatically becomes an intrusion and a threat to the existing power arrangements. It carries with it the menacing implication of displacement and disorganization of the status quo [1969:132].

To build people's organizations, he trained and dispatched a corps of organizers. The organizers served as stimulus and catalyst. Their task was to identify natural leaders (which Alinsky believed are present in every community) talk to them, get them interested in working together, and help them develop their native capacity. Organizers immersed themselves in community life to learn what was bothering the people and identify their self interests and to gain an understanding of individual and group experiences, habits, customs, and values. Since these traditions are often expressed through organizations, the organizer tried to work with all existing groups. As a result, the typical people's organization was built upon organizational representatives rather than individual members. They often use protest tactics as a means of getting attention and bargaining to achieve their demands.

1945 To 1960

World War II redirected the nation's energy away from domestic concerns, limited neighborhood activities, but didn't eliminate them entirely. As during World War I, civil defense activities had a neighborhood base in many cities. In Kansas City, Missouri City Manager L. P. Cookingham used civil defense workers to organize neighborhood associations and community councils around elementary schools for the purpose of dealing with wartime juvenile delinquency. After the war Kansas City continued to provide staff service for these organizations. In the postwar period other city governments began to pay attention to neighborhood organizations, particularly related to redevelopment and urban renewal projects.

REDEVELOPMENT

Philadelphia, 1945. World War II is over! It's time to return to domestic concerns, especially city slums and the lack of decent housing for low income people. A start was made during the thirties under the Works Progress Administration and the Housing Act of 1937, which had the triple objectives of slum clearance, job creation, and low rent housing. Even before the end of the war, civic organizations and the City Planning Commission began preparing postwar plans. Now the Pennsylvania legislature has passed a redevelopment law giving the city the power to acquire slum properties, relocate the occupants, demolish the buildings, and sell the land for new use. A Neighborhood Planning Conference has formed in South Philadelphia, and citizens are starting to draw up their own ideas on redevelopment. Residents and social workers in other neighborhoods are also talking about what redevelopment might mean for their neighborhoods.

As postwar planning for urban redevelopment proceeded, Philadelphia had experience with three different approaches (Slayton and Dewey, 1953:430-444). The first took place in the Southeast Central area through the initiative of the South Philadelphia Coordinating Council, an organization representing schools and welfare agencies. After a survey of conditions convinced the Council that the area was ripe for redevelopment, it set up a Neighborhood Planning Conference and invited every known organization to elect delegates. After a large initial meeting, committees were set up on various concerns and block meetings were held. Each committee had a consultant available, and the City Planning Commission sent staff to committee meetings. As the process unfolded, the Neighborhood Planning Conference added individual citizens to its memberships. When committee recommendations were ready, the Conference held public hearings and then adopted the package. The Planning Commission certified the area for redevelopment but then pretty much ignored the community's recommendations and instead produced its own redevelopment plan.

Second came the Popular Area in North Philadelphia. There the Planning Commission took charge and had its staff draw up the plan with almost no neighborhood consultation. After that, the Planning Commission had the Health and Welfare Council form a local committee to review the plan, but this group was unprepared to deal with the complicated proposal. An intent to present the plan to residents at a community meeting was aborted, and most of them heard about impending redevelopment from newspaper stories. Considerable opposition developed, but the city went ahead anyway.

The third approach occurred further north in the Temple Area where a field office of the Health and Welfare Council put together a committee representing home owners, tenants, schools, neighborhood associations, retail businesses, churches, industry, and real estate. Planning Commission staff then started their work with the neighborhood committee as a sounding board, and they made adjustments in response to committee comments. The committee took the proposed plan to the neighborhood at public meetings and gained greater support than occurred in the Popular Area. Yet, in Temple, as in the other two areas, the Planning Commission retained control of all planning decisions.

In this same era in Cleveland, neighborhood participation came through area councils set up under the aegis of Welfare Federation of Cleveland. The area councils differed in their makeup. Some had individual members, but most were primarily composed of such organizations as PTAs, churches, women's clubs, local business groups, veterans organizations, and block groups. Area workers of the Welfare Federation provided staff services. In redevelopment planning, the area councils took on the dual role of advising the Planning Commission and informing citizens about studies and proposals (Slayton and Dewey, 1953:455-163).

Chicago adopted a more structured approach through the formation of district planning boards under state enabling legislation. These boards were autonomous with their own staff and sources of funds. Within this framework the South Side Planning Board was controlled by representatives of two large institutions, Michael Reese Hospital and Illinois Institute of Technology, which were able to acquire sizable acreage through wholesale clearance and redevelopment. The Near West Side Planning Board (the site of Hull House, Chicago's oldest settlement house) had much greater representation of neighborhood people and tried to be more responsive to residents' needs. Nevertheless, eventually the City of Chicago adopted redevelopment plans which cleared the area to provide for a large campus for the University of Illinois, Chicago Circle (Slayton and Dewey, 1953:444-454).

SHIFT TO URBAN RENEWAL

Congress gave redevelopment a major boost by authorizing federal support through the Housing Act of 1949. After four years' experience the Eisenhower administration reviewed this clearance-oriented approach and offered recommendations for an urban renewal program with greater emphasis upon housing rehabilitation. Congress accepted this revision in the Housing Act of 1954, which also included a provision that a city must have a "workable program" in order to qualify for federal support. In its regulations the Housing and Home Finance Agency specified seven criteria for a workable program, including a requirement for citizen participation. This was a new element for a federally aided urban program.

In practice, strong citizen participation was the exception rather than the rule. Thus, when Gerda Lewis surveyed the first 91 cities with approved workable programs, she found that the communitywide advisory committee was the most widespread device to achieve citizen participation (1959:81). These committees were composed mostly of representatives from real estate, construction, downtown business, and citywide civic organizations. Minority groups had limited representation, and project areas almost none. In only a few places were there separate project area committees. The citywide advisory committees were limited to review of plans and policy recommendations, and in many places their role was vague.

Some localities did more, however. In a special demonstration project, the Housing Association of Metropolitan Boston placed organizers in various neighborhoods to boost resident involvement in urban renewal planning (Loring et al., 1957). In Philadelphia's Eastwick project, the Redevelopment Authority contracted with the Citizens Council on City Planning to provide an organizer. In Detroit, staff from the City Planning Commission assisted citizens in the Concord-Mack area, but eventually the strain of working for both the neighborhood and city hall became too great and the city withdrew its organizer. Likewise in Baltimore an attempt to build a strong community organizing component into a consolidated housing and urban renewal agency floundered on the issue of divided allegiance.

To be effective from the neighborhood perspective, staff support had to be independent. The difficulty was that there were scarcely any sources of funds for neighborhood organizers. The United Funds were too timid, or too committed to social services. Not many foundations were interested. Most neighborhoods, especially the poorer ones, didn't have or didn't try to raise the necessary revenues, though a few did. In Chicago, for instance, the Back-of-the-Yards Council had moved from protest to a neighborhood improvement orientation and supported its own staff (Millspaugh and Brenckenfeld, 1958), and the Hyde Park-Kenwood area near the University of Chicago hired a staff and developed its own renewal plan (Abrahamson, 1959; Rossi and Dentler, 1961; Cunningham, 1965).

EARLIER INITIATIVES REVISITED

The professionalization of social work, which began in the early days of the century, was complete by the 1950s. The profession had three major divisions: case work, group work, and community organization. The latter was considered "a process used by professional workers engaged in health and welfare planning" and a field of activity "occupied by agencies whose primary function is social planning, coordination, interpretation, or the joint financing of direct service agencies" (C. Murphy, 1960:186). Very few social workers engaged in grassroots neighborhood organizing.

The main exceptions were settlement houses and neighborhood centers. They, too, had become professional organizations, and their boards were composed mostly of nonresidents. Their programs had a heavy emphasis upon group work through building-centered activities, and they also provided case work services to individuals and multi-problem families. But they also studied neighborhood problems, helped organize block associations, neighborhood organizations, and tenant councils, and got involved in social action activities (Hillman, 1960).

Many schools continued to conduct after-hours programs, mainly recreation and adult education. But the zeal to serve as a focus for neighborhood organizations and to reach out to the surrounding community to encourage neighborhood action had long since disappeared from most community schools.

Civic associations were abundant, particularly in the suburbs and middle class city neighborhoods. In new subdivisions their natural history was to organize initially to deal with common problems the new homeowners had with the developer, next to take on the township, county, or whoever was supposed to provide local government services, then to sponsor recreation activities for children, and otherwise to become mostly dormant until some threat to tranquility arose. Property values and social homogeneity were major concerns. In 1944 24 separate associations in J. C. Nichols developments formed the Homes Association of the Country Club District for more efficient service operations. By 1960 there were approximately 500 homes associations around the country with maintenance responsibilities (Longhini and Mosence, 1978). Elsewhere in the postwar period citywide and county federations of civic associations formed, mainly to deal with city and county government on mutual concerns, usually property related issues.

In summation, various approaches to neighborhood organizing were in use during the 1950s, but in retrospect, considering the turmoil of the 1960s and higher levels of activism in the 1970s, they seem placid. Moreover, poor neighborhoods, which needed the benefits of concerted action the most, were the least organized.

Exercises

(1) For a neighborhood organization and a neighborhood institution in your city which originated before 1960, describe the social and political context of its origins, who the initiators were, why they acted, what the organization or institution did initially.

(2) Describe how the organization and institution have changed through the years.

(3) Read more about earlier neighborhood efforts, such those referred to in Chapters 4 and 8. Of particular interest are Addams (1910, 1930), Alinsky (1969), Dahir (1947), A. Davis (1967), Dillick (1953), Fisher (1981), Mumford (1961), and Perry (1929).

(4) For a more ambitious project, write a history of your neighborhood or some other neighborhood.

CHAPTER 9

THE TURBULENT SIXTIES AND AFTERMATH

The year 1961 marked a turning point in the history of neighborhood action. With the inauguration of a younger, activist president, John F. Kennedy, the national government became more assertive in promoting and supporting new approaches to old problems and in fostering community change. The civil rights movement was gaining momentum, and activists from other causes were becoming more numerous and more aggressive. Although social invention on neighborhood matters continued to occur mostly locally, as it had throughout American history, now federal agencies, national organizations, and some foundations were actively pushing the application of new approaches. These national entities began funding local organizers, sometimes sending them in from the outside. Social change was blowin' in the wind.

In many places, though, change efforts encountered mountains of resistance. The result was turbulence. Yet, change did occur. Out of this tumultuous decade came new ways which are still being absorbed and further developed in communities throughout the United States.

Federal Initiatives

COMMUNITY ACTION PROGRAM

> *Washington, D.C., November 23, 1964.* In a corner office on 19th Street, N.W., the director of the new Office of Economic Opportunity is signing the first grants of the Community Action Program. It brings to a conclusion a tragic and tumultuous year for Sargent Shriver: the assassination of his brother-in-law, President John F. Kennedy; the quick moves of the new president, Lyndon B. Johnson, to convert some half-developed ideas into a full-scale War on Poverty; Shriver's assignment to put the program together while retaining his position as director of the Peace Corps.
>
> What conflicting advice Shriver's has had about the Community Action Program. The Bureau of the Budget staff want it to be a coordinating mechanism. Daniel P. Moynihan from the Labor Department sees it as a political tool to make the program work to benefit the president, the

> Democratic party, and the poor. Richard Boone, who was with the President's Committee on Juvenile Delinquency, wants to keep the program out of the hands of social work professionals and local government bureaucrats and be certain that the poor themselves are involved. Jack Conway, with his mastery of coalition politics from his days with the United Auto Workers and the CIO, keeps talking about a three-legged stool involving established agencies, civic leadership groups, and representatives of the poor. At least these initial grants are in good hands: in Detroit and New Haven with loyal Democratic mayors, and elsewhere with local people the administration can count on.

These first Community Action grants launched the biggest infusion of money ever to initiate neighborhood-based services and community organizing. The Economic Opportunity Act of 1964 set the tone by requiring "maximum feasible participation of residents of areas and members of groups served." Furthermore, the time was ripe. The civil rights movement was starting to move in new directions, and here was a program easily adaptable to its needs. Moreover, there had been some promising starts in local programs funded by grants from the President's Committee on Juvenile Delinquency and the Ford Foundation's Gray Area Program.

The forerunners. The Juvenile Delinquency Committee got underway in 1961 with the new president's brother, Attorney General Robert F. Kennedy, in charge. Lacking its own appropriation for program grants, the President's Committee took effective control of demonstration funds authorized by the Juvenile Delinquency and Youth Crime Prevention Act of 1961. As a conceptual base, they adopted ideas advanced by Richard A. Cloward and Lloyd E. Ohlin in *Delinquency and Opportunity: A Theory of Delinquent Gangs.* These sociologists concluded that:

> services extending to delinquent individuals or groups cannot prevent the rise of delinquency among others. For delinquency is not, in the final analysis, a property of individual or even of subcultures; it is a property of the social systems in which these individuals and groups are enmeshed . . . The target of preventive action, then, should be . . . the social setting that gives rise to delinquency [1960:211].

Cloward and Ohlin felt that legitimate but functional structures must be developed to replace the traditional structures of slum communities. They were already trying to put their ideas into action through Mobilization for Youth on New York's Lower East Side, supported by planning funds from the National Institute of Mental Health. Mobilization for Youth received the first program grant from the President's Committee, 15 other communities got planning grants, and most of them later received program support.

In this same period the Ford Foundation was offering to make grants to communities willing and able to undertake comprehensive programs in "gray areas," a term used to describe inner city neighborhoods occupied by minority groups and other poor people. Oakland, New Haven, Boston, Philadelphia, and Washington, D.C., and also the North Carolina Fund, succeeded in attracting Ford support. The Oakland program was run by city government, but the others went to newly established, private nonprofit organizations. All the cities except Oakland also got juvenile delinquency planning grants (Marris and Rein, 1967; Thernstrom, 1969; R. Murphy, 1971).

Although there were many variations, these community efforts supported by the Ford Foundation and the President's Committee on Juvenile Delinquency shared a concern for opening opportunities for youth and disadvantaged adults. Neighboring organizing was a common ingredient, for they agreed with a Cloward-Ohlin premise that residents should be involved in programs serving them. As with the Chicago Area Project 30 years earlier (see Chapter 8), they hired a sizable corps of workers from within the neighborhood. Yet they were as traditional as mainstream social service agencies in the sense that their boards and principal staff were drawn from established organizations, city government, and reform-minded, upper-income individuals, not from the neighborhoods served. Their strength was program innovation, and many programs which later spread throughout the country under the War on Poverty — Head Start, Neighborhood Youth Corps, Legal Services, and many elements of the Community Action Program — were first developed in the gray area and juvenile delinquency programs.

CAAs. As community action agencies (CAAs) shaped up around the nation in 1965, 90 percent took the form of private nonprofit organizations. Most of the others were part of local government. Most places set up neighborhood units and also resident advisory committees to component programs, such as Head Start. The Office of Economic Opportunity encouraged at least one-third representation of the poor on CAA governing boards, and Congress made this part of the law in 1966. Most of the representatives of the poor were selected by neighborhood committees or area councils, which themselves were chosen by residents, usually in open meetings but occasionally by ballot or voting machines. Hiring of residents as professional aides and organizers was universal. Thus, what was unacceptable in most localities six years early in the Urban Renewal Program — hiring organizers with public funds — became commonplace under the Community Action Program. Such was the change that had occurred.

Virtually all of the urban community action agencies developed neighborhood-based program units. Social researchers at Brandeis University found that they were fulfilling several functions (1968:4):

- Decentralization of service programs.
- Formal and informal channels of communication between residents and the CAA.
- A formal framework for selecting members to the CAA board of directors.
- A channel for recruiting residents for jobs within the CAA.
- A means of stimulating and supporting neighborhood organizations to take action on poverty-related problems.

Although on the whole never as antiestablishment or radical as some opponents claimed, the Community Action Program constituted a sufficient challenge to existing ways of doing things that forces of reaction set in. A 1967 amendment required that one-third of the local community action boards must be local governmental officials but kept the requirement for at least one-third representation of the poor. This amendment also offered local government the option of taking over the program. Not many did, but the threat to do so, combined with tightened OEO regulations, constrained the more militant activities of community action agencies. Anyway, most of them were settling into mainly service operations after the initial thrust of organizing. They spent the bulk of their funds trying to remedy perceived defects of individuals rather than modifying institutions and social systems.

(For an overview of OEO programs, see Levitan, 1969. On local CAA experience, see Clark and Hopkins, 1969; Greenstone and Peterson, 1973; Hallman, 1967; Kramer, 1969; Lamb, 1975; Marshall, 1971; Rose, 1972; and Zurcher, 1970).

FURTHER FEDERAL ACTION

Model Cities. As a more peaceable alternative (it was hoped), the Johnson administration came up with the Model Cities Program. From the outset it was given to local government, but after months of internal debate, the Department of Housing and Urban Development, which ran the program, issued a guideline stating that

> there must be some form of organization structure, existing or newly established, which embodies neighborhood residents in the process of policy and program planning and program implementation and operation. The leadership for that structure must consist of persons whom neighborhood residents accept as representing their interests (HUD, 1967).

When James L. Sundquist and David W. Davis of the Brookings Institution surveyed the 75 cities receiving the first round of Model Cities grants (1969:96), they discovered five organizational patterns: ranging from resident control to a unicameral city agency at the extremes and in between three variations of a bicameral arrangement, one variety city hall oriented, one neighborhood oriented, and the third unified. Thus, four out of five patterns provided roles for neighborhood residents in policy formulation. However, in another analysis Sherry Arnstein indicated that only 15 of the 75 cities achieved some "significant degree of power-sharing with residents, and in all but one of these places, it occurred because of angry citizen demand, not city initiative" (1969:222).

After the Model Cities Program had operated for five years, George J. Washnis documented its effects in a sample of eight cities. He found staff dominance in four cities, resident-dominance in two, and parity in two. He concluded that on the whole the citizen participation process had positive effects on program products. Beyond that, he wrote (1974:59), "citizen participation has sought and found new resident leadership, forced some important changes in government, brought democracy and decision-making closer to the people, and involved at least some of the poor in the actual working of government." (Also see HUD, 1968; Powledge, 1970; Model Cities Service Center, 1971; Marshall Kaplan, Gans & Kahn, 1973; Hinckley, 1977; S. Weissman, 1978.)

Urban renewal. Community Action and Model Cities experience had an impact upon the older Urban Renewal Program, as residents demanded and in many localities achieved greater participation in urban renewal planning. In 1968 HUD issued a regulation mandating project area committees in all renewal projects involving housing rehabilitation (by then, most of them did), and permitting federal funds to be spent for that purpose. Many cities provided staff services to these committees, and some were allowed to hire their own staff (National Urban League, 1973; also see, Colburn, 1963; H. Kaplan, 1973; J. Wilson, 1963; C. Davies, 1966; Burke, 1966; and Keyes, 1969).

Other programs. Requirements for citizen participation spread to other federal-aid programs: education, social services, health, employment and training, economic development, transportation, water resources management, and many more. In 1978 the Advisory Commission on Intergovernmental Relations (ACIR) tallied 155 federal grant programs with statutes or regulations mandating citizen participation. This amounted to more than one-fourth of all grant programs, but more significantly they accounted for over 80 percent of federal grant expenditures. ACIR reported that "Among the major modes of participation, over one-half of the programs — 89 — require boards or committees reflecting the public in various ways in their membership. Fifty-five programs mandate public hearings while in 114 programs other types of citizen participation are specified, i.e., public meetings, workshops, and review and consultations." Of the 89 programs requiring citizen boards or committees, 24 provided for some decision-making power while the other 65 were advisory only (ACIR, 1979:112-113).

Not all of these federal programs dealt with neighborhoods, and for those which did, the degree of neighborhood involvement varied from slight to substantial. But quite a number did provide meaningful roles for neighborhood residents, and some also channeled funds to neighborhood corporations for their own operations. Compared to the degree of neighborhood involvement in public programs occurring in the 1950s, federally mandated citizen participation marked a noteworthy advance. Compared to what civil rights advocates and community organizers were demanding, the requirements fell considerably short of the activists' aspirations. But as we'll notice in Part III when we review neighborhood roles in a number of program areas, federal regulations and money have significantly aided increased neighborhood participation.

Citizen Activists in the 1960s

CIVIL RIGHTS MOVEMENT

Mississippi, June 1966. Five years ago white rioters tried to block James H. Meredith from enrolling as the first black student at the University of Mississippi, and federal marshals had to come to protect him. Since then civil rights advocates in the United States have achieved significant gains, particularly two major legislative victories in Washington with enactment of the Civil Rights Act of 1964 and the Voting Rights Act of 1965. This summer Meredith decided to test how much things have changed in

> Mississippi by walking from Memphis to Jackson. On the first day he was ambushed and wounded. Now people have come from all over the nation to continue the march. They've camped at the edge of town this evening and are gathered to hear Dr. Martin Luther King, Jr. A warm-up speaker asks: "What do you want?" "Freedom!" is the reply. "When do you want it?" "Now!" But a group from the Student Nonviolent Coordinating Committee (SNCC) is gathered around Stokely Carmichael, and they shout "black power" as their answer to the first question.

Though united in a fundamental commitment to equality, the civil rights movement was pluralistic in its organizations, which reflected different histories and different emphases. The National Association for the Advancement of Colored People (NAACP) came into being in 1910 as an embodiment of political and legal activism advocated by W.E.B. DuBois; Roy Wilkins took the reins as executive director in 1955. The National Urban League was established in 1911 to help southern migrants adjust to urban living conditions in the North, thus expressing Booker T. Washington's view that Negroes should concentrate on economic progress; but when Whitney M. Young, Jr., became executive director in 1961, the organization became more activist. The Congress for Racial Equality (CORE) formed in 1941 as an offshoot of the pacifist Fellowship of Reconciliation to use techniques of nonviolent direct action to fight racial discrimination, and James Farmer became national director in 1961. The Southern Christian Leadership Conference, headed by Martin Luther King, Jr., got going in 1957 after the Montgomery, Alabama bus boycott of 1955-56 and other southern endeavors led by black clergymen. Carmichael's organization, the Student Nonviolent Coordinating Committee (SNCC), originated in the South in 1960 and spread to northern cities; it was the most vocally militant organization (National Advisory Commission on Civil Disorders, 1968:216-236).

These organizations united with organized labor in sponsoring the mammoth 1963 March on Washington, initiated by A. Phillip Randolph, president of the International Brotherhood of Pullman Workers, who had used the threat of a similar march in 1941 to persuade President Franklin D. Roosevelt to issue an executive order requiring fair employment practices in war industries. The civil rights movement had substantial white support from churches, national labor leaders, liberal politicians, and other segments. This combined array of forces achieved legislative success in enactment of the Civil Rights Act of 1964, which outlawed discrimination in public accommodations (the target of many of the direct action campaigns), and the Voting Rights Act of 1965.

Because of racial discrimination in housing, membership in the various civil rights organization was concentrated in predominantly black neighborhoods (though there were white members drawn from elsewhere), but the movement wasn't particularly territorially oriented. This changed during the 1960s for two reasons. First, civil disorder which wracked numerous cities from 1963 through 1968 revealed the seething unrest and exposed the deplorable living conditions of black urban ghettos. Second, the growing demand for black power, which gained public expression during the Meredith March, was soon translated into demands for community control of institutions and programs in black neighborhoods.

Stokely Carmichael and Charles V. Hamilton articulated this viewpoint in *Black Power: The Politics of Liberation in America* (1967:4):

> The adoption of the concept of Black Power is one of the most legitimate and healthy developments in American politics and race relations in our time. . . . It is a call for black people in this country to unite, to recognize their heritage, to build a sense of community. It is a call to reject the racist institutions and values of this society.

They described the social dynamite of the urban ghettoes, talked about the need for new organizational forms, and argued that (1967:166) "We must begin to think of the black community as a base of organization to control institutions in that community."

Already the Community Action Program had created some opportunities for neighborhood control. Saul Alinsky had turned his talents to organizing black communities in Chicago, Rochester, and elsewhere. Then in the late 1960s some of the organizing zeal of the civil rights movement was directed toward the cause of community control.

WELFARE RIGHTS

Many of the urban community action agencies opened storefront offices in inner city neighborhoods in order to reach the poor more effectively. As the people came in with their problems, the staff realized that lack of money was the root of many difficulties, that numerous residents were eligible for public assistance but weren't receiving it, and that some recipients weren't getting what they were entitled to or were encountering other difficulties with the public welfare department. This brought the CAA workers into an advocacy relationship with welfare agencies. As an ally they gain support from the OEO-funded legal services, and CAA organizers put together groups of welfare recipients so that they could be their own advocates. In 1966 George A. Wiley left a staff position with CORE to open a Poverty/Rights Action Center in Washington. Within a year he and others brought together some 350 persons representing 200 welfare groups in 70 cities from 26 states in a national meeting. From this meeting came the National Welfare Rights Organization, which in turn fostered more local welfare rights groups. Although not neighborhood organizations in a territorial sense, most of them were based in poor neighborhoods and functioned as militant advocates in dealing with neighborhood outposts as well as central headquarters of the welfare department (Piven and Cloward, 1971:287-338).

YOUTHFUL WHITE ACTIVISTS

> *Newark, New Jersey, June 1964.* Led by Tom Hayden, 13 members of the Students for a Democratic Society (SDS) have settled in the lower Clinton Hill section in the South Ward. Working with some local residents in this mostly black neighborhood, they have formed the Newark Community Union Project for the purpose of building a social protest movement. The students are going door-to-door and talking to people on the street to find out the issues disturbing them. They want to enlist residents to help in the organizing, and soon they intend to call some neighborhood meetings.

This project lasted for about three years. Seven white students and 15 residents gave almost full time, another 25 residents were intensively involved for extended periods, and more than 100 others participated in public demonstrations, rent strikes, meetings, and other organizational activities. They focused on three main issues: housing, a traffic light, and an electoral contest. Housing action took the form of rent strikes, which resulted in some evictions but no notable improvements in housing conditions. The traffic light they wanted was for a particularly dangerous intersection on a major thoroughfare. Residents went to city hall, blocked traffic, but never succeeded in penetrating the bureaucratic maze. Having been thwarted by the politicians, they joined a coalition of dissident blacks, Puerto Ricans, and civil-rights-oriented whites to back three candidates for the New Jersey Assembly on a third party ticket. But the candidates gained only five percent of the vote. By the end of 1967 the Newark Community Project ceased functioning with scarcely any successes to show (Parenti, 1970). SDS organizers moved on to other communities to organize against the Viet Nam war and prepare for massive demonstration protests at the 1968 Democratic national convention in Chicago.

Elsewhere, though, others of their generation performed more successfully in local organizing under other auspices. For instance, Volunteers in Service to America (VISTA), a War on Poverty program, attracted thousands of middle class whites into volunteer service in both rural and urban poverty areas. Others got jobs with community action agencies, and those with legal training became lawyers for Neighborhood Legal Services.

In this same period a number of young city planners concluded that their profession was captive to real estate and other economic interests to the disadvantage of people in poor neighborhoods. To serve the poor more effectively, they would become advocate planners.

> The planner as advocate would plead for his own and his client's view of the good society. The advocate planner would be more than a provider of information, an analyst of current trends, a stimulator of future conditions, and a detailer of means. In addition to carrying out these necessary parts of planning, he would be a *proponent* of specific substantive solutions [Davidoff, 1965:333].

However, as advocate planners entered poor neighborhoods, they began to realize that they themselves weren't value free and had to guard against becoming another manipulator of the poor (Peattie, 1968). They also found that their services weren't always welcomed by those they sought to serve (M. Kaplan, 1969). Those who worked from staff positions within city planning departments had to deal with conflicts between what neighborhoods wanted and overall city policies going in the opposite direction (Needleman and Needleman, 1974; also see Mann, 1969; Hallman, 1970:173-176; Blecher, 1972; Mazziotti, 1974; Heskins, 1980).

CIVIL DISORDER

Rioting and looting was another kind of citizen activism of the 1960s, particularly in black ghettoes. Starting with several disturbances in 1963 and repeated in following years, the outbreak of civil disorder peaked during the first nine months of 1967 with 167 occurrences: eight major ones in terms of violence, property damage, and loss of life; 33 serious but not major; and 123 minor in character. The National Advisory Commission on Civil Disorders, appointed by President Johnson to determine causes and solutions, identified a variety of complex and interacting factors. The commission concluded that "of these, the most fundamental is the racial attitude and behavior of white Americans toward black Americans. Race prejudice has shaped our history decisively in the past; it now threatens to do so again. White racism is essentially responsible for the explosive mixture which has been accumulating in our cities since the end of World War II" (1968:203).

The commission came up with a host of recommendations for social and economic programs to respond to conditions or urban ghettoes. A central proposal was the following (1968:16):

> City governments need new and more vital channels of communication to the residents of the ghetto; they need to improve their capacity to respond effectively to community needs before they become community grievances; and they need to improve opportunity for meaningful involvement of ghetto residents in shaping policies and programs which affect the community.

Among the specific recommendations were neighborhood action task forces, neighborhood city halls, and multiservice centers. In the same year the National Commission on Urban Problems (1968:350), also appointed by President Johnson, recommended decentralization of municipal services to the neighborhood level and efforts to establish channels of communication with neighborhood residents.

Neighborhood Control

During the last half of the 1960s the call for black power was translated locally into a demand for community control. (For a summary of arguments, see Altshuler, 1970). The demand was strongest for control of agencies and programs which minority communities felt weren't serving them properly or justly, particularly the police and public education. In response some police departments enlarged their community relations staff, opened storefront offices, set up neighborhood advisory committees, and assigned police to specific neighborhoods (see Chapter 12), but no neighborhood gained control over the police. Large school systems set up study groups to consider ways of achieving greater community participation, but only in New York and Detroit moved toward greater community control by establishing elected boards for subdistricts within the city (see Chapter 13).

In this period several proposals for neighborhood government were forthcoming from white advocates, including Milton Kotler (1969), Joseph F. Zimmerman (1972), myself (1974) and David Morris and

Karl Hess (1975). The proposals had two sets of major themes (Naparstek, 1976). One set emphasized participatory democracy, local liberty, and self-rule. The other stressed administrative effectiveness and the orderly division of responsibilities among several tiers of local government. The themes came together by insisting that neighborhood government, controlled by the residents, would provide services more responsive to individual needs.

No city has adopted full-fledged neighborhood government, but some degree of community control has developed in many localities. It takes its simplest forms in voluntary, self-help activities such as those we chronicle in Chapter 11. In an old tradition, residents decide something needs doing, and they do it. This kind of neighborhood self-help burgeoned during the 1970s. In addition, two organizational species emerged which moved closer to neighborhood control: officially recognized neighborhood councils with advisory roles on city policy matters and neighborhood corporations with operational responsibilities.

NEIGHBORHOOD COUNCILS

Dayton, Ohio, December 1971. The returns are in, and 149 residents have been elected to Dayton's five neighborhood priority boards. They join the 21 elected members of the Model Cities Planning Council in being responsible for making recommendations for allocation of $5.2 million in federal funds under Planned Variations of the Model Cities Program. The Model Cities Planning Council has been playing this role in West Dayton since 1969 when residents negotiated an "equal partnership" agreement with the City Commission. Last year the city set up the five priority boards and divided $200,000 of city funds among them for special projects which they designed. The City Commission also refers all proposed zoning changes and other issues affecting their neighborhoods to the priority boards.

During the first half of the 1970s a number of other cities set up neighborhood councils, defined as broad-based organizations of residents, usually elected, with official recognition by city government and assigned advisory roles on matters affecting their neighborhoods (Hallman, 1977a:4). There had been earlier experience in Kansas City, Missouri, as noted in the previous chapter. And in the 1950s the president of Manhattan Borough in New York City appointed community planning boards to advise on capital programming and other city planning officials. A 1961 charter amendment applied this concept citywide, and subsequent amendments have added to their roles, though they have remained appointed bodies (now simply called community boards).

In 1967 the Advisory Commission on Intergovernmental Relations drew up model state legislation authorizing cities and counties to set up neighborhood subunits with advisory powers (ACIR, 1967). During the next several years bills based on this model were introduced in the legislatures of Oklahoma and Minnesota but not adopted. However, new city charters in Honolulu, Pittsburgh, the District of Columbia, New York, and Newton, Massachusetts authorized neighborhood councils. Similar recommendations came from charter commissions in Los Angeles, Boston, Chicago, Detroit, and Rochester, but these proposed charters went down to defeat over other issues. City councils passed ordinance or resolutions creating neighborhood councils or setting up a process to officially recognize existing neighborhood associations in such places as Birmingham, St. Paul, Wichita, Anchorage, Eugene, Salem, and Portland, Oregon. In other locales less formal systems developed to give quasi-official recognition to neighborhood associations. In 1971 the California legislature adopted an act authorizing counties to establish municipal advisory councils in unincorporated areas, and by 1977 there were 27 of them in eight counties (Fuller, 1977). The Ohio Commission on Local Government Services (1974) encouraged cities to recognize neighborhood subunits. (For samples of local charter provisions and ordinances, see Center for Governmental Studies, 1974; National Association of Neighborhoods, 1978; Andersen et al., 1979. On the operation of neighborhood councils, see Hallman, 1977a and Chapter 16).

NEIGHBORHOOD CORPORATIONS

During this same period the number of private nonprofit corporations controlled by neighborhood residents grew rapidly. Many got started with funding from New Frontier and Great Society programs: Manpower Development and Training, Juvenile Delinquency, Community Action, Special Impact, Model Cities. They continued to grow under revised legislation adopted during the Nixon and Ford administrations, including the Comprehensive Employment and Training Act of 1973 and the Housing and Community Development Act of 1974. The Ford Foundation, other foundations, a few labor unions, and some national church boards provided financial support. We'll run into many varieties of neighborhood corporations in Part III, especially Chapters 12 to 15. They have gained control of many different activities, operating as private nonprofit organizations rather than as neighborhood governments.

Movements in the 1970s

Citizen activism flourished in many forms during the 1970s and was expressed in a variety of social movements (Boyte, 1980). Some were constituency oriented, such as radical and ethnic minorities, women, seniors. Others focused on issues, such as the environment, energy, product safety and other citizen concerns, economic justice, war and peace. Among them, the neighborhood movement gained many new local coalitions and a set of national organizations.

ANTI-REDLINING

> *Chicago, Illinois, 1971.* The lobby of this northwest neighborhood bank is unusually crowded, even for a Saturday. Long lines of people are waiting at each window to put $1.00 in their accounts and then withdraw 50¢ in pennies. It is the final tactic in a campaign which began when a Polish homeowner was turned down for a mortgage loan and a Puerto Rican resident was rejected for a business loan, even though both were well qualified. A committee from the Northwest Community Organization called on bank officers and forced an admission that the community itself was the problem. The citizens demanded a mortgage pool for the community and the right to review anyone rejected for loan. When the bank turned down this proposal, neighborhood activists handed out leaflets on two successive Saturdays, informing depositors that the bank was redlining the neighborhood. This made no impact, so they resorted to the $1.00 and 50¢ tactic. Unexpectedly an older woman has thrown her 50 pennies on the floor, and the outburst has so rattled the bank officials that they are arranging an immediate meeting for the group with the chairman and the president of the bank.

At this meeting the group won their demands, plus a $1,000 donation for their organization (National Peoples Action, 1975).

Other neighborhood organizations in Chicago were also concerned about redlining, so named because allegedly lending institutions draw a line around neighborhoods where they won't make loans, or demand very strict terms. One of them was the Organization for a Better Austin, which had been established by an Alinsky-trained organizer (Bailey, 1974). Gale Cincotta was a leader who emerged from that organization, and she joined with others to convene a National Housing Conference in Chicago in March 1972. It drew 2,000 delegates from 74 cities in 36 states. Out of it came a national membership association, National Peoples Action on Housing, and a support organization known as the National Training and Information Center.

In the Chicago area 13 community groups formed the Metropolitan Area Housing Alliance, which led a campaign that succeeded in getting the Commission of Savings and Loan Associations for Illinois to adopt an anti-redlining regulation in January 1974. Turning their attention nationally, this force, supported by grassroots organizations in many other cities and some Washington-based national organizations, pushed until they got Congress to adopt the Federal Home Mortgage Disclosure Act of 1975, the Community Reinvestment Act of 1977, and legislation authorizing a presidentially appointed National Commission on Neighborhoods (National Peoples Action, 1975).

TECHNICAL ASSISTANCE PROVIDERS AND RESEARCH INSTITUTES

The growth of National Peoples Action and other national neighborhood organizations, discussed below, was greatly facilitated by the staff and network connections of national and regional technical assistance organizations which came into being in the late 1960s and the 1970s.

Jack Conway, who had moved from the AFL-CIO's Industrial Union Department to head the OEO Community Action Program, organized the Center for Community Change in 1968 to provide technical assistance to community development corporations. Monsignor Geno Baroni, working from a base in the U.S. Catholic Conference, set up the National Center for Urban Ethnic Affairs (NCUEA) in 1970 to provide assistance to business people and residents of predominantly white ethnic neighborhoods. NCUEA had particularly close ties with Gale Cincotta and her National Training and Information Center. Milton Kotler, who had been based at the Institute for Policy Study, established the Institute for Neighborhood Studies in 1971 and began making contact with grassroots neighborhood organizations, particularly in the northeast and midwest. In 1969 I organized the Center for Governmental Studies (later renamed Civic Action Institute) to conduct research on neighborhood decentralization, particularly emphasizing what local governments were doing. The National Council of La Raza, the National Urban League, the National Urban Coalition, and the Support Center are among a long list of other technical assistance providers which have worked with neighborhood and community based organizations during the last 15 years.

A number of these organizations received Ford Foundation support. They also got grants from smaller foundations and negotiated technical assistance contracts with federal agencies. Mostly they were initiated by activists working at the national level rather than emanating from the grassroots, but their network connections helped bring together local people who then formed national membership associations.

Also around the country were various centers training and dispatching community organizers, such as the New England Training Center for Community Organizers in Providence, the Industrial Areas Foundation and the Midwest Academy in Chicago, and Organize Training Center in San Francisco. In many cities coalitions formed, usually with a political action orientation, but some of them gained technical assistance staff.

A number of small research institutes were set up by individuals having a major interest in neighborhoods, such as John McClaughry's Institute for Liberty and Community, Nelson Rosenbaum's Center for Responsive Governance, Robert Woodson's National Center for Enterprise Development, Donald Warren's Community Effectiveness Institute, and Robert Hawkins' Sequoia Institute. Elinor Ostrom put together a network called the Neighborhood Organization Research Group, which has merged with a newer, broader University Consortium for Neighborhood Research and Development. The citations in this book reflect the diversified, scholarly research which has been occurring on neighborhood topics. (For an analysis of neighborhood research, see P. Olson, 1982.)

NATIONAL ASSOCIATIONS

Another phenomenon has been the creation of national, neighborhood-oriented membership association. Thus, in 1969 some people from racially integrated neighborhoods in a number of cities got together for a conference and formed National Neighbors, Inc. as an organization dedicated to fostering and maintaining integrated neighborhoods. We have already noted how National Peoples Action got underway in 1972. In the spring of 1975 Milton Kotler called a meeting in Washington out of which came the Alliance for Neighborhood Government, which after several semiannual meetings changed its name to the National Association of Neighborhoods (NAN). In my studies of neighborhood decentralization I discovered that a number of cities, unbeknownst to one another, had established recognized neighborhood councils, so I invited practitioners from those cities to join in a workshop in Kansas City, Missouri in May 1976. They decided to meet again the following year as the National Conference on Neighborhood Councils, which in 1980 changed its name to Neighborhoods, USA (NUSA). Some women who were active in NAN formed the National Congress of Neighborhood Women in 1979 with headquarters in Brooklyn.

In addition, some issue-oriented national and state organizations have undertaken neighborhood organizing to build support for their own agendas. One of these is ACORN (Association of Community Organizations for Reform Now), which evolved from an original organization in Arkansas established in 1970 by Wade Rahtke, who came out of the welfare rights organizing movement. Using young, idealistic, low-paid organizers, ACORN mobilizes grassroots support on such issues as utility rates, taxes, and abandoned houses and in some places gets into direct political action by supporting candidates for public office. It has a number of state affiliates, such as Carolina Action, and local branches (Boyte, 1980:93-97).

There are also some neighborhood-related associations composed of particular kinds of local organizations and personnel, such as the National Congress for Community Economic Development made up of representatives of community development corporations, the National Community Design Centers Directors Association, the Neighborhood Arts Program National Organizing Committee, United Neighborhood Centers of America (formerly the National Federation of Settlements and Neighborhood Centers), and the Neighborhood Planning Division of the American Planning Association. The pluralism of neighborhood interests is reflected in this diversity.

NATIONAL NEIGHBORHOOD COALITION

After one aborted attempt, representatives from the Washington-based technical assistance providers and national neighborhood organizations have formed the National Neighborhood Coalition as a vehicle for information sharing and joint action on common concerns, with particular attention to federal legislation and administrative policy. It grew out of a smaller, ad hoc coalition which formed in 1979 to muster support for several pieces of legislation related to neighborhood interests. Most of the organizations mentioned in the two preceding sections are members along with a variety of other national organizations which have neighborhoods among their varied interests, such as the League of Women Voters, National Trust for Historic Preservation, and the Corporation for Enterprise Development. Its purposes include information sharing, working for increased resources for poor and disadvantaged neighborhoods, and serving as a forum for analysis of proposed federal laws and regulations.

REMNANTS AND ENDURING APPROACHES

Locally in the 1980s neighborhoods have many remnants of the War on Poverty and other Great Society initiatives. Quite a number of community action agencies have persisted, now pretty much settled into a social services routine with scarcely any zeal for citizen organizing. Neighborhood corporations of many varieties are now the cutting edge of federally assisted efforts. Head Start has become well accepted in most places and continues to provide advisory roles for parents. The demand for neighborhood control of public schools has lessened, but an interest in parental participation has continued and community school programs are fairly widespread (see Chapter 13).

Settlement houses and neighborhood centers remain as useful service organizations, and some of them maintain an interest in organizing and advocacy. Other kinds of neighborhood service centers are commonplace (see Chapters 12 and 13). Civic associations remain active in the suburbs and middle class city neighborhoods, continuing to focus mainly on property values and social preservation. The number of neighborhood associations in older, low- and moderate- income neighborhoods has grown in the past 20 years; they vary in their activism but in some cities mobilize to seek benefits for their neighborhoods. There are now an estimated 35,000 "homes" or "community" associations, oriented toward use and maintenance matters for clusters of homes or condominium units (Dowden, 1980).

In sum, residents in a majority of American neighborhoods can find an organizational outlet for their civic interest, if they choose to participate. Or they can organize one, using a set of proven techniques, which we'll look at in the next chapter.

Analysis

This virtual catalogue of neighborhood-related activities is far from complete, but it touches the main feature of the neighborhood movement during the past 25 years. Several comments can provide an analytical perspective.

POLITICAL ORIENTATION

Many of the neighborhood initiatives of the past 100 years have been directed toward the needs of poorer neighborhoods, both in neighborhood organizing and service delivery. Local neighborhood coalitions, national neighborhood organizations, technical assistance providers, and training centers tend to focus on the needs of the have-nots, and as Alinsky added, the have-a-little, want-mores (1971: 19). This has placed the neighborhood movement in the liberal band of the conventional political spectrum. Not "radical left," for varied attempts to use neighborhoods as the base for organizing the proleteriat have never had enduring success. And not "radical right" either, for in spite of articulating values of self-initiative and local control, most of those espousing this cause are more attuned to the have-lots and have no interest in the needs of poor neighborhoods.

Yet, there is a strong conservative streak in most neighborhood organizations, certainly in the property-oriented, suburban civic associations and also in organizations in low- and moderate-income neighborhoods. That's because a neighborhood is a fixed place, a collection of buildings and land, mostly privately owned. Owners want to preserve property values, and many tenants would like to own or in some other way to "get a piece of the pie." Many neighborhood organizations also want to gain some form of control over particular neighborhood endeavors. They may use protests techniques as a means to this end, but once they are in charge of operations, or have some other kind of stake in neighborhood programs and facilities, they become part of the system and settle into more conservative ways. That's why some organizers argue against neighborhood organizations getting into direct operations.

Thus, public officials and other establishment leaders and journalists who perceive of the neighborhood movement as radical are misled by the rhetoric of some advocates. They don't grasp the mainstream focus of most neighborhood organizations: on having a good place to live and raise children, a chance for residents to find jobs, become homeowners, have good schools, be safe and secure. To the extent that a maldistribution of resources prevents some neighborhoods from achieving these goals, the holders of excessive wealth and power might not want to see neighborhood demands for fairness articulated. But most Americans should have no fear of the demands of people who want the same things for their neighborhoods that they themselves desire.

DIVERSITY

Some observers look at the diversity of the neighborhood movement and perceive excessive fragmentation and duplication. They remark that much more could be accomplished through consolidation. Yet, as James Cunningham has pointed out, pluralism is the nature of any social movement; otherwise, it wouldn't be a movement. Not that working together is undesirable, for unity proves strength in dealing with strong and sometimes centralized opponents. Yet, a major asset of the neighborhood movement is the initiatives taken by many strong-minded individuals and organizations dedicated to their primary concerns. They come together on shared issues but retain their own identity. This residual individuality may be a weakness, but it is a weakness of a free society, whose basic freedom offers even greater strengths. (For another critique, see Goering, 1979.)

NEED FOR GREATER SUPPORT

In a book concentrating on neighborhoods, we should be cautious in asserting that the movement is very powerful. Although there have been gains during the past two decades, many neighborhoods don't have truly effective organizations and lower income neighborhoods face a constant struggle to preserve themselves and improve living conditions.

A major cause is the lack of strong and enduring public commitment to ending conditions of poverty in America. Furthermore, many public officials, civic leaders, and scholars are skeptical that neighborhood still has meaning in this age of population mobility, high technology, and nationally dominated economic forces, and accordingly they won't acknowledge the importance of neighborhood action. There are many persons who concentrate on other interests and causes without paying any attention to neighborhood impact. What they do may be neutral toward neighborhoods, supportive, or outright harmful, though not necessarily intentionally.

For instance, a study of resource allocation in Oakland, California, found that some neighborhoods benefited more than others from decisions affecting schools, libraries, and streets, not purposely but rather as a byproduct of policies made for other reasons arising from professional and bureaucratic commitments (Levy et al, 1974). Similarly lenders' preoccupation with maximizing profits and returns for their depositors and overcautiousness in seeking safe investments led to redlining practices, though racial and class prejudice may also be a factor.

Whether inattention to neighborhood needs stems from lack of interest and ignorance of harmful impact or from pursuit of other goals contrary to neighborhood interests, neighborhood activists still have occasions for using protest techniques to gain attention to their grievances, show how they are harmed, prove that they have a large following, and get policies changed. But after protest they must have organizational capacity to negotiate with public agencies and private interests and to take on direct operations if desired.

The recession of the 1980s and the substantial withdrawal of federal support for neighborhood programs have been particularly harmful to lower income neighborhoods (Palmer and Sawhill, 1982; Piven and Cloward, 1982). Neighborhood organizations have had to work hard to survive, but most of them are proving their durability (Cohen, 1983). Among residents of poor neighborhoods, gaining sufficient income to pay the rent and buy groceries usually rates a higher priority than political action to gain neighborhood control over service delivery. Among city governments the maintenance of basic services is a greater concern than initiating new experiments with little city halls and other neighborhood projects (though there is a positive connection between productivity and decentralized delivery for some services). Consequently compared to the 1960s and the 1970s, neighborhood innovations requiring public expenditures have slowed. But as the next part of the book will show, neighborhood action in the 1980s continues strongly in many fields of service with the participation of many, diverse actors.

Exercises

(1) Find a remnant of the Community Action Program, Model Cities Program, or another of the Great Society programs functioning in a neighborhood in your community. Describe how it has changed over the years.

(2) Look for a neighborhood organization which started during the last ten years. Who organized it? Why? What does it do? What segment of the neighborhood movement is it a part of?

(3) Is there a neighborhood coalition in your community, either citywide or among only a few neighborhoods? Who are its members? What are its main issues? How does it pursue those issues? Does it put aside any issues because they could cause disunity? What difficulties does it encounter in staying together?

(4) Why do social movements arise? How do they keep going? Why do they seem to die out? Apply these answers to the neighborhood movement.

CHAPTER 10
STRATEGIES AND METHODS OF NEIGHBORHOOD ORGANIZING

When we look back at one hundred years' experience, we observe that people have used a variety of methods to initiate neighborhood action. Methods range from provision of direct services to advocacy. In between are coordinating and advisory mechanisms. Some programs have used combined approaches. Table 10.1 categorizes the methods used by the activities discussed in the two previous chapters.

As this table also summarizes, initiation came from a variety of sources: private and public agencies (in the last 30 years often from federal programs), developers, concerned individuals from outside the neighborhood, and the residents themselves. Residents gained control over some, but not all, of the activities initiated by outsiders. Two trends are evident during the last 20 years: greater amount of neighborhood control and heavier federal involvement in funding neighborhood activities. The two are related, but in complex ways. As citizen activists, working both locally and nationally to increase the power and influence of neighborhoods, have been able to get federal funds channeled to neighborhoods and regulations oriented toward greater citizen participation. They have found allies in the human rights movement and here and there within the ranks of government.

Strategic Choices

If we look at approaches to neighborhood action in the 1980s from residents' perspective, we find two sets of strategic choices. The first choice is whether to seek direct control over neighborhood operations or to concentrate upon influencing what others do. The second choice is whether to stick to what's happening in the neighborhood or also to deal with persons, agencies, and institutions in the broader arena who affect neighborhood life. In exercising these choices, residents function as a political community and can select among alternatives we've considered in Chapter 5. These are summarized in Table 10.2.

TABLE 10.1 Methods of Neighborhood Action, by Selected Activities, 1870 to Present

	Emphasis					
Period, Activity	*Operations*	*Advocacy*	*Advisory*	*Coordination*	*Initiator*	*Neighborhood Control*
1870–1918						
Charity organizations	P				Private agency	No
Settlements	P	s			Concerned outsiders	No
School centers	P	s			Public agency	Partial
Homeowners associations	P	s			Developer	Yes
Civic associations		P			Residents	Yes
1919–1960						
Area coordinating councils				P	Public agency	No
Chicago area project	P	s			Private agency	Partial
People's organizations		P			Concerned outsiders	Yes
Urban renewal committees			P		Public agency [a]	No
1961 to present						
Community Action Program	P	P			Private agency [a,b]	Some components
Model Cities Program	P	s	P	s	Public agency [a]	Shared in some localities
Other federal programs		s	P		Public agency [a]	Usually not
Citizen activists		P			Residents, concerned outsiders	Yes
Neighborhood corporations	P	s			Residents [c]	Yes
Decentralized administration	P			s	Public agency	Usually not
Neighborhood councils		s	P		City agency	Yes
Self-help associations	P	s			Residents	Yes

a. Part of a federal program.
b. Public agency in a few places.
c. Some with federal support.
NOTE: P = primary emphasis; s = secondary emphasis

Table 10.2
Choices of Strategies, Focus, and Methods for Neighborhood Action

Strategy	*Focus*	*Methods*
Influence	Electoral politics	Voting Representation Political parties
	Interest group activities	Advocacy Protest
	Structured participation	Hearings Advisory committees Boards and councils Other methods
Control	Traditional voluntary activities	Self-help Human services
	Government services and programs	Municipal services Public education Housing development
	General purpose government	Suburban municipality Neighborhood government within central city
	Public/private sectors	Economic development

If neighborhood residents want to influence decision makers, they can get involved in electoral politics, act as an interest group, or take advantage of opportunities for structured citizen participation. Methods of electoral action include mobilization of voters, involvement in party politics, and seeking direct representation in policymaking bodies. Interest group activities encompass various methods of advocacy, including protest tactics. Among the methods of citizen participation are hearings, advisory committees, neighborhood councils, and various techniques of communications and task-oriented planning.

If neighborhood residents want direct control of programs and services, they can embark upon self-help activities and and take on the kind of services traditionally handled by voluntary agencies, especially in human services. They can try to take over services ordinarily handled by municipal agencies through contracts or other form of delegation and can get involved in government-funded housing and economic development programs. They can seek control of public schools within the neighborhood. Or, they may want to form a general-purpose government with its own governing body, revenue sources, and a broad range of administrative responsibilities.

The choice between these main strategies — influence or control — is affected by the geographic focus of action. In dealing with the broader arena residents are more likely to use influence tactics because one part has difficulty controlling the whole. However, they may seek the right to affirm or veto decisions affecting their neighborhood, or gain representation on broader governing bodies. Where residents seek control, they are more likely to concentrate upon services, programs, and development projects occurring within the neighborhood. But they may prefer to gain a policy voice or substantial influence without directly administering neighborhood operations. Their choice affects the organizational form they use, for one type of organization is required for advocacy and protest, another for program operations.

In practice, a mixture of strategies is common, sometimes adopted by a single neighborhood organization or more frequently pursued by several sets of neighborhood actors and organizations.

At first glance, to control seems stronger than to influence, but the difference should be judged by content and impact, not rhetoric. Thus, a neighborhood might exert tremendous effort to gain control of a minor municipal service while paying little heed to city economic development policies seriously affecting neighborhood life. Or, a neighborhood organization might use all its energies to rehabilitate a few houses while neglecting policies of public agencies and private lending institutions impacting all the neighborhood's housing. But on the other hand, protestors may feel they are a vital force but gain only a few token concessions. Or, residents who appear at public hearings, lobby council members, and serve on official advisory committees may think they are influential when in reality actual control remains firmly in the hands of appointed officials. So, strategies should be evaluated by results yielded in particular circumstances.

To carry out any of these strategies, a neighborhood must first be organized. In the remainder of this chapter we'll examine techniques of neighborhood organizing, and especially the use of interest group tactics to achieve influence. We'll mention tactics of gaining control but reserve for Part III how neighborhood control unfolds in major program areas.

Neighborhood Organizing

All neighborhoods have social networks linking residents with one another, and most have many varieties of organizations. Many neighborhoods have associations consciously looking out for neighborhood interests, and other organizations with other aims are drawn into interest group activities as issues arise. But sometimes new issues emerge which no existing organization is equipped to handle. And some neighborhoods lack a viable organization to deal with issues affecting the neighborhood. Therefore, in a variety of circumstances organizing is necessary if neighborhood interests are to be cared for. From decades of experience a panoply of neighborhood organizing techniques has developed (Alinsky, 1969, 1971; Warren and Warren, 1977; Miller, 1979; Hallman and Wegener, 1980; Kahn, 1982; Cunningham and Kotler, 1983).

INITIATORS

To form a new organization or transform an old one, somebody has to take initiative. In neighborhood organizing the initiators might be residents, nonresidents regularly employed in the neighborhood, or an outsider coming in specifically to organize. Resident initiators might already be serving as leaders in other organizations, such as church, PTA, youth groups, a social club, or they might be riled up by some new issue or inspired by a persuasive call to action. Initiation might come from persons in institutional roles, such as school principals, clergy, settlement house workers, community development staff, or business proprietors. An organizer might be sent in by an outside organization or requested by a group of residents. There can be a combination of initiators.

Motivation among resident initiators varies. Some are activated by a particular crisis. Some are angry and tired of being pushed around (Kahn, 1982:23). One study asked leaders of 11 neighborhood organizations in Indianapolis to explain their motivations and found four categories of responses: (1) a desire to fulfill perceived civic duties (39 percent); (2) a devotion to the neighborhood and desire to serve its interests (39 percent); (3) obligations of friendship (15 percent); and a desire to protect property values (6 percent) (Rich, 1980a:576-577). A study of neighborhood residents participating in decentralization experiments in New York and New Haven determined that they became active primarily because of a desire to preserve or improve their home neighborhood, an opportunity for a new mode of involvement, and natural exuberance for activism (Yates, 1973:88-89). In neither study was a desire for economic gain or social status a strong motivating factor (perhaps because small-scale neighborhood activities don't offer such rewards), but participants did gain feelings of personal esteem and pride in their accomplishments. Nevertheless, there may be some potential indirect benefits by advancing one's professional or political career or making business contacts.

Initiation by persons associated with neighborhood-based agencies and institutions tends to be an extension of their occupational role. For instance, the school principal acting because of a concern for the environment in which education occurs. Community development staff responsible for citizen participation. Clergy who have a social concern for neighborhood conditions.

Outside organizations sending organizers into neighborhoods usually concentrate on low income, working class, or minority areas. They do this for humanitarian reasons, to garner support for particular programs, or to gain political allies (Taub et al., 1977). The humanitarian concern is to equip residents with the capacity to work for their own interests. Program orientation relates to a desire to increase citizen participation and enlist support. Political motivation is to mobilize allies on issues and causes championed by the outside organization. Such causes range from radical to conservative, such as fighting banks and utilities on the one hand to resisting racial integration on the other. (For the effects of outside assistance, see Cooper, 1980.)

Regardless of who the initiator is, organizing comes easiest when focused upon particular issues of great concern to the residents or upon a crisis. Among these are perceived threats to personal safety or property values; concern for neighborhood children, youth, or the elderly; or rumored plans of changes in land use, such as redevelopment, highway construction, or industrial expansion; increases in taxes, property assessments, or rents; population changes perceived as undesirable; or cutbacks in popular programs. Sometimes issues are submerged and have to be brought to the surface. It is also possible to foster organizations to deal with neighborhood planning, program development, and other longer term activities, but this usually requires more intensive effort and is likely to gain a lower level of participation.

PREPARATION

Any kind of organizing not involving a sudden crisis requires careful preparation. Where it is done by an outside organizer, that person needs time to understand the neighborhood, explore issues of concern, and identify actual and potential leaders. But even indigenous initiators need time to get things together.

To community organizations, Saul D. Alinsky offered this advice (1969:64):

> The building of a People's Organization can be done only by the people themselves. The only way the people can express themselves is through their leaders. By *their* leaders we mean those persons whom the local people define and look up to as leaders. . . .
>
> These indigenous leaders are in a very true sense the real representatives of the people of the community. They have earned their position of leadership among the people and are accepted as leaders.

Leadership patterns vary in different neighborhoods. An analysis of neighborhoods in the Detroit area identified three main types of local leaders (Warren and Warren, 1977:63-64): the activist, who has a reputation for taking action but is not a member of an organization; officers of local organizations; and opinion leaders, persons who are approached frequently for advice or knowledge about a particular problem. Some neighborhoods are loaded with organizations while others are less organized. All of them are also composed of overlapping social networks, as we discussed in Chapter 3, and persons at nodal points function as informal leaders. An organizer can talk to residents to find out whom they look upon as leaders, and can visit with shopkeepers, agency personnel, and other service workers for their perceptions.

While doing this the organizer can also gain a better understanding of residents' primary concerns and work with them to identify issues around which to organize. Issues, Si Kahn has taught (1982:90), "are problems that people feel strongly about and want to do something about. But in order to be a good issue, a problem must have a solution that can be achieved by people working together." Furthermore, he explained, a good issue affects and involves a lot of people, is strongly felt, clear and simple, unites the people, and builds the organization.

Another part of preparation is forming a core group of leaders who will take responsibility for making preliminary decisions on the form of organization, calling initial meetings, and defining the agenda. Later the organizing process will move from this small nucleus to a wider base of participation.

FIRST MEETINGS

Decisions on the form of organization determines the nature of the initial meetings (Hallman and Wegener, 1980:9-12). Some neighborhoods start with block organizing and only later form a neighborhood federation of block groups. In this case, the first meetings occur in people's homes, one block at a time. A block organizer (often a volunteer from that block, though maybe trained by a community organization staff) goes door to door inviting people. The first meeting combines discussion of one or two issues with informal socializing and concludes with a decision on when and where to meet again. After a meeting or two, participants elect officers who will keep the group going. Usually it is a good idea to initiate an action project in their block, such as crime watch, cleanup, or beautification, so that people will know that they are achieving results. From this base they can expand to neighborhood activities and action on broader issues (Wandersman et al., 1981).

An alternative is to form a neighborhood association from the outset. After a period of preparation, the core group announces the initial meeting and invites participation. Experienced organizers differ on whether to try for a small or large meeting. Some prefer to start with a small, cohesive group which can agree upon an agenda and quickly move into action. Others want as much involvement as soon as possible and therefore stage a large, mass meeting. In advance the core group works out the agenda, perhaps drafts some resolutions, and is ready to present a proposal for formally organizing. If the meeting flows smoothly, a bylaws committee and a nominating committee is appointed. But sometimes residents don't want to be rushed, so it takes another meeting or two of discussion before setting up a permanent organization.

ORGANIZATIONAL FORMS

According to James V. Cunningham and Milton Kotler, neighborhood organizations have three types of structures (1983:13):

> Most common is the membership structure where individual persons or families join and represent only themselves. There is also the coalition structure where the members are organizations and institutions. A third, newer kind of structure, is found in organizations tied closely to government and politics. Structure in these is built upon representatives of sub-neighborhood territories elected by their local constituencies. Combination structures are rare, but possible.

Neighborhood associations formed through initiative of resident leaders tend to be of the first type. Organizers trained in the Alinsky tradition favor the coalition structure, and block federations also take this form. Neighborhood councils organized with official recognition are usually of the representative variety. Sometimes they will have representatives from both geographic subareas and organizations.

Most neighborhood organizations set up some kind of executive committee or board of directors, chosen by the membership. Selection occurs in three ways, sometimes in combination: at an annual congress or convention; by direct voting at polling places or by mail with either only signed-up members or all residents eligible to vote; or designation by local organizations. Annual membership meetings also adopt broad policy guidelines. A few neighborhood organizations are governed by a town meeting or assembly, which meets regularly to make policy decisions, but they usually have some kind of executive council to handle organizational affairs between assembly meetings.

SUSTAINING AN ORGANIZATION

Once established, a neighborhood organization needs to find ways to sustain support and active participation over the long haul. This is particularly true for an organization initiated in a crisis atmosphere. Choice of tactics, which we'll get to in a moment, affects how an organization maintains interest, but other factors are also important.

Careful and thorough fact-finding is an asset to any issue-oriented organization, whether its style is confrontational or cooperative problem solving. Some neighborhood organizations have staff who can do this, but most of them must rely upon volunteers for this task. They might include retired persons, housespouses and other nonemployed residents, students from the neighborhood and nearby colleges and universities. Many cities have legal services and citywide advocacy organizations which provide assistance on particular issues. Under the federal Community Reinvestment Act, passed after pressure from National People's Action and allies (see Chapter 9), data are available on mortgage lending patterns of financial institutions. The U.S. Census contains neighborhood statistics. City planning departments and United Way planning bodies produce useful data. And there are lots of other places where a neighborhood organization can dig up information.

Recruiting volunteers and developing new leaders are perennial challenges to neighborhood organizations. If they don't, they risk dominance by a leadership clique and gradual diminution of participation as leaders and other volunteers burn out. The place to start is by defining precise tasks volunteers can perform and actively recruiting people to fill these roles. Most volunteers need training and supervision, whether by another volunteer or a staff person. From the core of volunteers new leaders emerge. They, too, need training in leadership responsibilities and methods. Some cities have leadership training institutes serving many grassroots organizations. (For more ideas, see Biaggi, 1978.)

Public recognition of volunteers and elected leaders acknowledges their contributions, builds up their esteem, helps publicize the work of the organization, and encourages other residents to volunteer. This might be done at the annual meeting, a fundraising event, or a special banquet for this purpose.

Staff support is helpful, though many neighborhood organizations get by without any. Ordinarily this is a paid position, maybe only part time or a consultant who works for several organizations. Sometimes a dedicated volunteer can fill this role. Some neighborhood organizations receive staff support from a public agency or a citywide voluntary organization. When there is staff, the relationship among elected leaders, the board of directors, and staff has to be worked out carefully so that confusion and competition over roles doesn't detract from a smooth operation (Flanagan, 1981).

To have staff and to meet other expenses, neighborhood organizations must raise funds. This can be done through membership dues (both individual and organizational), fundraising events, door-to-door canvassing, ads in the annual convention program, and many other techniques of grassroots fundraising (Flanagan, 1982; Johnsen et al., 1982). Some cities with official neighborhood councils make modest annual grants to them, and community development agencies sometimes fund project area committees and other citizen participation bodies. Community foundations and certain national organizations, such as the Campaign for Human Development supported by the Roman Catholic Church, make grants to neighborhood organizations. Although many of these outside funding sources respect the independence of organizations they support, there are situations where the one paying for the piper wants to call the tune.

A desirable, but too often neglected, task for sustaining an organization is evaluation. This process is easier if the organization has a clear statement of its objectives, detailing precisely what it wants to achieve within a given time. Then a committee, or the whole board of directors, can periodically assess progress and problems in meeting the objectives. This helps to keep the organization on target, or to realize that its purposes are shifting and may require adjustments in tactics. (See

Weiss, 1972; Spiegel, 1977; Rosener, 1978; Cunningham, 1979; and N. Rosenbaum, 1980.)

TACTICS

A neighborhood organization has a choice of many tactics. Si Kahn has offered a long list of options (1982:187):

> Common tactics include: strikes, boycotts, picket lines, sit-ins, public hearings, confrontations, press conferences, paid advertising, visits to public officials, actions, mass demonstrations, marches, petitions, letters, exposes, lobbying, leaflets, prayer services, silent vigils, civil disobedience, rallies, and legal action. In a long or drawn-out campaign several of these tactics may be used. Some tactics may be used twice or even several times.

The selection of tactics relates to the overall strategies of the organization, such as whether it is seeking to influence or gain control and, if to influence, whether it is concentrating on electoral politics, interest group activities, structured citizen participation, or a combination. Another factor is whether the organizing process stresses conflict or cooperation. The school of organizing in the Saul Alinsky tradition emphasizes conflict, stemming from a belief that low-income people and minority groups have real grievances against the ruling elite and they can be mobilized and gain victory only through aggressive actions (Alinsky, 1969, 1971; Miller, 1979). Another school insists that solutions to neighborhood problems require joint action by government, private enterprise, voluntary organizations, and residents. Therefore, organizing should be done in a spirit which builds foundations for future cooperation (Hallman and Wegener, 1980). (For a range of cases, see Bailey, 1974; Marciniak, 1977; Lancourt, 1979; Cunningham and Kotler, 1983.)

Regardless of approach, the neighborhood organization needs to decide exactly what it wants to achieve, whom it has to deal with to accomplish its objectives, and how best to reach these persons. Demands should be precise but yet flexible enough so that if another way is found to achieve the objective it can be accepted. Having determined whom it wants to influence, the organization must determine how to communicate its desires. Alternatives are letters, telephone calls, visits, appearances at scheduled hearings, and, for persons not wanting to receive the group, picket lines and sit-ins. Sometimes an indirect approach is appropriate, such as working through other persons who have access to decision makers or using mass media to gain attention.

If the organization enters the arena of electoral politics, it can get its members on precinct and ward committees, get delegates elected to nominating conventions, hold candidate forums, initiate registration and get-out-the-vote campaigns, provide transportation on election day, and monitor the polls and vote counting.

Although dismissed by some radical organizers as futile, many neighborhood organizations take advantage of the processes of structured citizen participation. They get hold of key documents, study them, and offer comments. They turn out their troops for public hearings, serve on advisory boards and task forces where they can make their case, develop counterplans, and use their entry into the decision-making process as an opportunity to bargain on behalf of neighborhood demands. And they monitor program implementation (Eisenberg, 1981; also see references on citizen participation in Chapter 5).

Skill in negotiations is useful whether neighborhood representatives get to the bargaining table through official processes or in response to protest actions. There, absolute demands usually have to give way to accommodation and compromise. It's not always an easy transition for militant neighborhood leaders to make, but may be necessary to achieve results.

Sometimes neighborhoods realize that they share interests with other neighborhoods, or with other constituencies, such as tenants, racial and ethnic minorities, environmentalists, homeowners, taxpayers, and others. They then form coalitions to unite their strengths in dealing with whomever they consider to be their common adversary (Spiegel, 1981). Coalitions may function temporarily around a particular issue, or they may be more enduring. A local neighborhood coalition may involve every neighborhood in the city, or it may take in only certain ones which perceive a common interest, such as lower- and moderate-income areas. To the extent that neighborhoods are competitive with one another, matters of their competition (such as share of public funds) might be excluded from the coalition agenda. Or, they might use the coalition as a forum to negotiate an agreement of a fair share for all neighborhoods, thereby strengthening their position as they contend with other interests trying to tap the same funds.

The varied tactics we've been considering can be used either to influence policies of decision makers in government and the private sector or to gain neighborhood control of specific operations. When the latter happens, a shift must occur in how the neighborhood organization proceeds. Then it will have staff to direct, programs to manage, and funds to spend, and it will be accountable for results. While the board of an action organization is usually composed of very active volunteers who do most of the work, a board of a neighborhood corporation concentrates more on policy matters and paid staff handle operations. In Chapter 16 we'll look more closely at organizational issues arising under neighborhood control, after we've studied actual neighborhood operations in a number of fields in the intervening chapters.

Philosophical Issues

Neighborhood organizing raises two important philosophical issues: how one views human nature and the relationship of means and ends.

HUMAN NATURE

Experienced organizers realize that appeal to immediate self-interest is the easiest way to organize. For some, this is founded upon a philosophical viewpoint that people are motivated only by self-interest. Others, though, point out that people also have a concern for others: parents for their children, grown children for elderly parents, residents for neighborhood youth, citizens for the good of the whole city. They draw upon people's concern for others as well as self-interest in their organizing activities.

This debate is also found in academic circles. Conventional economists erect their theories upon a view that people are basically motivated by rational self-interest, and "public choice" political-economists extend this belief to an analysis of political interest groups (M. Olson, 1965; Bish and Ostrom, 1973; O'Brien, 1975). For them, the challenge in community organizing is to find incentives appealing to self-interest which will activate citizens to work for the public goods and services to be enjoyed by all members of a community. In contrast, other scholars hold the view that people are also sometimes nonrational, are social beings, and are motivated by concern for others as well as for self. Accordingly, neighborhood mobilization should draw upon these interrelated aspects of human nature (Henig, 1982).

My own experience has taught me that we humans combine self-interest and concern for others. You can find this social concern in kinship relationships, friendships, and mutual aid among neighbors. Within neighborhoods people give money, food, and clothing to those in need, and volunteers work many unpaid hours on community endeavors. Indeed, the neighborhood as a social community is rooted in personal relationships and our concern for one another. When given a chance, this same trait carries over to a neighborhood's relationship with other neighborhoods and with the broader community. But never without self-interest also present. For each of us individually and for every neighborhood, self-seeking and self-giving occur intermingled.

MEANS AND END

In *Rules for Radicals* (1971:24-47) Saul Alinsky claimed that the perennial issue of whether the end justifies the means is wrongly phrased. Instead he maintained that the real question in the ethics of means and ends should be, "Does this *particular* end justify this *particular* means? In this vein, he insisted:

> The man of action views the issue of means and ends in pragmatic and strategic terms. He has no other problem; he thinks only in his actual resources and the possibilities of various choices of action. He asks of ends only whether they are achievable and worth the cost; of means, only whether they will work.

Mahatma Gandhi offered a different viewpoint, which I share. In *Hind Swaraj* (India Home Rule), he stated (1934:95): "The means may be likened to a seed, the end to a tree; and there is just the same inviolable connection between the means and the end as there is between the seed and the tree." That is to say, the end is contained in the means. Therefore, no just end can come out of an unjust means.

Martin Luther King, Jr., another apostle of nonviolence, addressed this issue in explaining why we should love our enemies (1963:37-38):

> Darkness cannot drive out darkness; only light can do that. Hate cannot drive out hate; only love can do that. Hate multiplies hate, violence multiplies violence, and toughness multiplies toughness in a descending spiral of destruction. . . .
>
> Another reason why we must love our enemies is that hate scars the soul and distorts the personality. Mindful that hate is an evil and dangerous force, we too often think of what it does to the person hated. . . . Hate is just as injurious to the person who hates. . . .
>
> A third reason why we should love our enemies is that love is the only force capable of transforming an enemy into a friend.

Neither King nor Gandhi, though, stood by passively accepting injustice. Rather they led vigorous campaigns of community and political change. They used many of the tactics we have considered.

They recognized that direct confrontation was often necessary, but they did it in a manner which kept open the door for amicable settlement of grievances. They were willing to negotiate. Their nonviolent methods, which respected their adversaries while opposing them, made it more likely that both sides could come to agreement. And so it can be in neighborhood organizing.

Exercises

(1) For a neighborhood organization, or some other organization to which you belong, identify the initiators. What were their motives? What are their rewards? How many are still around? If there have been changes, how do you account for them?
(2) For any leadership position you hold, and for other leaders of voluntary organizations you participate in, list motives and rewards.
(3) Attend a meeting of a voluntary organization and observe how it functions. Who are the leaders? What is their style? Do they allow participation? How do they move the meeting to decisions? How do they handle disagreements?
(4) For your neighborhood, or another one you are acquainted with, list the residents' major concerns. Who needs to act to resolve these issues? What are the best tactics for getting them to act?
(5) How do you perceive human nature regarding self-seeking and self-giving?
(6) What do you consider to be the relationship of means and ends?

PART III

NEIGHBORHOOD-BASED ACTION

CHAPTER 11

SELF-HELP ACTIVITIES

Having traced the history of neighborhood action and explored the strategies and techniques of neighborhood organizing, we are now ready to examine the kinds of activities, services, and projects which are now going on in American neighborhoods. We start by looking at some small-scale, self-help activities, and then we devote successive chapters to larger operations dealing with physical preservation and revitalization, municipal and human services delivery, economic activities, and governance. Self-help cuts across these latter fields but deserves separate consideration because of certain common ingredients.

Neighborhood self-help refers to activities ordinarily initiated by residents, carried out mostly by volunteers, and achieved with relatively small amounts of money and supplies. The adverbial modifiers are necessary in this definition because occasionally outside organizers initiate self-help activities, and sometimes resident volunteers receive modest staff support. Some self-help activities evolve into more heavily funded service operations, and then their nature changes.

From an enormous variety of neighborhood self-help occurring nowadays, we'll sample activities dealing with safety and security, housing and physical environment, neighbors caring for others, and communal events.[1]

Safety and Security

People are concerned about their personal safety and the security of their homes and possessions. In addition to what they do individually, residents get together to carry out crime prevention activities and other protective measures.

BLOCK WATCH AND PATROLS

During the late 1960s and early 1970s block watch activities and neighborhood patrols began in a number of cities (Washnis, 1976). Neighborhood leaders were the initiators in some places, such as West Philadelphia. Elsewhere the police department took the initiative but enlisted neighborhood support, as in Los Angeles. Sometimes a citizens crime prevention committee and the police department worked together, as occurred in Oakland. Methods developed by innovators spread by word-of-mouth. Then the U.S. Congress appropriated funds to the Law Enforcement Assistance Administration (LEAA) in the Department of Justice for a federal grant program, aimed at extending community crime prevention. From this experience, a well-developed body of techniques has emerged (New York, 1978:12-22; Wegener, 1979; Skogan and Maxfield, 1981; Baltimore, 1982:72-78; Cook and Roehl, 1983).

The essence of block watch and patrols is people looking out for one another. To do this, they must be organized. Therefore, block organizing is the heart of this approach, calling into play techniques we considered in the previous chapter. Close cooperation with the police is also an essential ingredient because residents aren't trying to take the law into their own hands.

Block watchers keep an eye out for suspicious activities, such as anyone forcibly entering a home, business, or car; unusual noise, such as breaking glass or screaming; anyone with a weapon; a stranger offering young children candy or money; adults loitering around schools, playgrounds, or secluded areas; persons who seem to have no purpose in the neighborhood. When something suspicious is happening, the block watcher calls the police, describes exactly what she or he has seen, the location, license number and other identification, and remains on the phone in case the police need additional information. Many blocks put up signs to indicate that they are part of the neighborhood watch with the intention of deterring crime.

Patrols are the next level of organization in community crime prevention. They consist of unarmed persons, usually in pairs, walking or driving around neighborhood. They might have a civilian-band radio to keep in touch with patrol headquarters in someone's home, or walkers might knock on a neighbor's door and ask to use their telephone. Either way, they call the police to apprehend alleged offenders but don't attempt this themselves. Most of all they enable the neighborhood to reassert control of the streets so that residents can feel safe to come out, even after dark.

Some patrols and individuals carry *whistles or freon horns* and blow them when they are in trouble or observe criminal acts in process. Hearing the whistle or horn, members of the block group call the police.

OTHER MEASURES

Through *Operation Identification* residents use an electric pen to put identification numbers (such as social security number) on televisions, stereo sets, and other appliances. If stolen and recovered, they can get back their property. By placing decals on their windows indicating that they participate in Operation Identification, they may deter burglary.

Some neighborhoods provide *escort service* to elderly persons traveling to and from senior centers, clinics, or other places to protect them from muggers. Parents form patrols for their children. In New York a union of building maintenance workers, who work at night, organized a buddy system to pair workers living near one another (Washnis, 1976:72-75).

Neighborhood businesses have *buddy alarm systems* which they can silently activate during a holdup so that the "buddy" merchant can call the police. Merchants also sponsor patrols on commercial streets (New York, 1978:23).

FIRE PREVENTION

Protection from fire is another neighborhood concern. This starts with each household eliminating fire hazards and having an escape plan in case of fire. *Volunteer fire wardens* can inspect buildings to identify safety hazards, such as congested hallways, overloaded circuits, defective electric cords, and improperly stored combustible material. They can encourage the installation of smoke alarms. Neighborhood associations can buy alarms for households who can't afford them (New York, 1978:24-25).

In neighborhoods with abandoned buildings arson may be a problem, with fires set by vandals, vagrants, or at the instigation of building owners seeking insurance payments. If a neighborhood has a crime prevention patrol, it can add *arson watch* to its responsibilities. Or, a special antiarson patrol can focus on vacant buildings, construction sites, schools, and other sites where arson might occur (Baltimore, 1982:85; Cook and Roehl, 1983).

MEDIATION

Not everyone is neighborly in neighborhoods. Family tensions break out of the home. Disputes arise among neighbors over noise, uncontrolled pets, fences, property maintenance, and other matters. Adults are bothered by youths. Although some of these things might lead to acts requiring police intervention, many such disputes could be resolved without adjudication. To do this, a few localities have instituted *neighborhood mediation boards* and other means of dispute resolution (Baltimore, 1982:88).

Housing and Physical Environment

Preservation and improvement of property and the physical environment is another focus of neighborhood self-help. Housing particularly receives a lot of attention.

HOUSING

Although owners have basic responsibility for housing maintenance and local governments sponsor programs to deal with housing preservation, improvement, and new construction, neighborhood residents frequently embark upon small-scale, self-help activities to improve housing quality.

Paint-up, fix-up is one of the most common self-help approaches to housing (Uplift, 1974:220-223; People Power, 1979:122-128). It is particularly applicable in neighborhoods with many homeowners, although absentee owners can also be drawn in. As part of an organized effort, the property owners voluntarily agree to paint, repair roofs, porches, and stairs, and make other external improvements. Residents pitch in and help one another, and volunteers assist elderly homeowners. There may be free paint available, paid by a small city grant, the neighborhood association's own funds, or a donation from a citywide organization. In some localities city inspectors issue compliance orders under the housing code to bring about improvements by absentee owners who aren't cooperating.

Some neighborhoods have *tool lending libraries*, formed to enable residents to share tools which they need only occasionally (Sampler, 1979:23-27; Baltimore, 1982:49). This might include both hand and power tools for carpentry, plumbing, electrical work, masonry, lawn care, and landscaping. Tools might be acquired by purchase or donation by residents and members. There can be a membership fee to pay all or part of the costs. There must be a space to store the tools, a system for lending, and volunteers or a small staff to run the tool library. Disposable items, such as sandpaper, caulking, and other supplies can be available for sale.

Skill exchange is something which happens informally in most neighborhoods as neighbors help one another with tasks they are good at. This can be organized into a bartering system with credits earned and drawn upon. In this way, a person may be repaid by someone other than the one for whom he or she did work. Housing repairs might be only one of many skills and services in a neighborhood bartering program (Simon, 1979; Fletcher and Fawcett, 1979; Tobin, 1980; Tobin and Fletcher, 1981).

Energy conservation offers other opportunities for neighborhood self-help (Hallman and Goldwasser, 1980; Baltimore, 1982:113-122). This can consist of education on the merits and methods of saving energy, cooperation with local utilities on energy audits, and workshops on best way of weatherizing and other energy-saving improvements. Neighborhoods can get together for bulk purchase of insulation and heating oil. Youth and other volunteers can help weatherize the homes of the elderly.

ENVIRONMENT

Neighborhood cleanup is another common activity (New York, 1978:27-32; Baltimore, 1982:29-43). Residents select a date when they will work together to clean rubbish in backyard, alleys, and vacant lots. They make arrangements with the city to close off the street and to provide a special refuse collection truck. This may also be an opportunity for residents to get rid of old appliances and other bulky objects which the regular pick-up won't handle. As part of the cleanup campaign, residents might sweep sidewalks and collect litter from street gutters. Many blocks have a party at the end of the day to celebrate their clean block.

Some of the material collected during cleanup day can be *recycled.* This means that it must be separated, properly packaged, and loaded for delivery to a recycling center. Sometimes neighborhoods relate to larger community-based recycling operations (CAI, 1979b; Fresno County Economic Opportunities Commission, 1979; Mulligan and Powell, 1979).

While cleanup programs might occur once or twice a year, residents can work together year long to keep their neighborhood clean and sanitary. They can have regular *sidewalk sweeps.* Higher density neighborhoods may need litter baskets, a campaign to get residents and passersby to use them, and a commitment from the city to empty them. If rats are a problem, residents can make a concerted effort to close harborage sites, keep garbage out of reach, and work with the city's *rat eradication* program (Baltimore, 1982:34-35).

OPEN SPACE

Most neighborhoods have open spaces which can be a community asset if properly developed, maintained, and utilized (New York, 1978:41-57; Baltimore, 1982:52-58).

Often streets and sidewalks compose the greatest amount of open space. They can be enhanced through *tree planting and care.* Some cities will plant the trees. In other places a city forester can advise what species do best in particular locations, and the block group or neighborhood association can then hire a private contractor (who may need a city permit). Once planted, each tree can be adopted by a resident, who agrees to water it, weed around it, see that supports are tight, ward off vandals, and report any problems to the city or contractor.

Vacant lots can be turned into *totlots, elderlots,* or *gardens.* Ordinarily the permission of the owner is required, whether it be an individual, a corporation, or the city, but where a neighborhood can't find the owner or penetrate the city's bureaucratic maze, residents sometimes claim squatter's rights. They clear the lot of debris, bring in equipment and landscaping material, open it for use, take care of it, and watch over it to prevent misuse. (A more formal approach is to form a land trust, which we'll discuss in chapter 15.)

Building walls facing onto outdoor spaces can be enhanced by *mural painting.* Some neighborhoods also *paint fireplugs* and work with merchants to *harmonize signs* as a means of beautification.

In these days of tight city budgets, a neighborhood association might *adopt a park* to supplement what the city can do in maintenance. If not a whole park, a city square or a small piece of land at street intersections. Also, in springtime block groups can purchase flowers wholesale for individual households to plant on their own property, including in window boxes (Bush-Brown, 1969).

Caring Neighbors

Mutual aid is a natural part of neighborhood life. It occurs through numerous actions of individuals who know their neighbors' needs and care for them. Yet, some needs go unmet because they aren't well known or nobody is in a position to respond. Another thrust of neighborhood self-help, therefore, is to undertake activities of "caring neighbors" in an organized manner. This can serve all ages in many different ways.

CHILDREN

Between the family, which has fundamental responsibility for child care, and child welfare agencies and institutions are informal and small-scale neighborhood-based activities (New York, 1978:73-77; Baltimore, 1982:90-92).

Day-care for young children is strictly regulated in most localities, especially if more than a smaller number (say, five) are served on a premises. However, usually a person may take care of a few children in a residence. While home day-care providers and parents often make direct connections, a neighborhood association can act as a facilitator by sponsoring surveys and helping to match potential providers with those in need of day-care services.

Baby-sitting cooperatives are formed so that parents can swap services. This can be part-time day care or evening and weekend baby-sitting as parents go out for leisure time activities or shopping, to worship or meetings. Sitting can be done with several children in the sitter's home, or the sitter can go the child's or children's residence. When they sit, participants earn credits which they can draw upon when they want a sitter.

In addition to direct services, a neighborhood association, or a special committee, can serve as a *children's advocate.* One model for this is found in Massachusetts, which has 43 local councils for children receiving staff support through a statewide program. Where they function in cities, they serve community districts of several neighborhoods, but a single neighborhood could undertake the same activities. Their task is to determine extent of services to children within the locality, provide information and referral services, determine service gaps, make recommendations for new activities, and monitor and evaluate existing children's services. Among other results, councils of children have documented lack of response to child abuse and neglect and brought about stronger community-based efforts (CAI, 1979a).

YOUTH

Millions of Americans serve as volunteer youth leaders through PTAs, boy and girl scouts, recreation programs, social service agencies, churches, and many other auspices. A significant portion of these youth programs is neighborhood-based and functions in neighborhoods along the entire income spectrum. Here we want to give particular attention to a couple of activities aimed at youth who are having difficulty in school, job finding, and staying out of trouble with the law.

Homework help and tutorial programs can be organized in neighborhood schools, churches, community centers, and apartment buildings. They can mobilize both adults and youth who are doing well in school to help those who are having difficulty. This kind of activity requires a coordinator, a regular schedule, and close cooperation with the school. It is likely to require donation of reference books, reading and writing materials, and small calculators, and perhaps access to a computer (New York, 1978:71).

Youth counseling and employment opportunities are other needs which neighborhood organizations can meet. One notably successful example is the House of Umoja in Philadelphia where Sister Falaka Fattah and David Fattah have taken youth gang members into their house to live. The Fattahs tutor and counsel them, prepare them for job interviews, help them find jobs, and provide love and discipline. In Chester, Pennsylvania Youth-in-Action started working with youth gangs from a base in the home of its founder, Mrs. Tommie Lee Jones. The program worked with gang leaders to eliminate gang warfare, and then got the members into self-rehabilitation through peer counseling and community service. In other communities neighborhood-initiated efforts have reached school dropouts and other youths considered intractable by professionals and have succeeded in redirecting their lives (Woodson, 1982b).

THE AGING

Most older persons prefer to live independently as long as they have the health and financial resources to do so. At the same time, they want opportunities to participate in community activities. As they become more frail, some of them need various kinds of support in order to continue living in their own homes and getting to places. Some have relatives and long-time friends available, but many older persons lack these connections and can benefit from assistance provided by nearby volunteers. Conversely, neighborhood organizations can provide a wide variety of services for the aging (New York, 1978:79-91; CAI, 1979d, Ehrlich, 1979; Baltimore, 1982:94).

Senior centers offer a place where older persons can come together for education, recreation, meals, fellowship, and sometimes health care. They also offer a base for outreach to the homebound elderly. Home services can also be handled by a neighborhood operation which doesn't have building activities.

The place to start in serving the homebound is an *outreach program*, designed to survey the neighborhood to find out who they are, where they live, and what their needs are. This is necessary because many elderly persons are out of sight and forgotten. Next comes *home visits* on a regular basis to offer friendship, delivery of books and magazines, serve as liaison to social service agencies, and on occasion act as a troubleshooter and advocate. There can be *daily telephone contacts*, initiated either by the older person or by a volunteer for the purpose of reassuring her or him that somebody cares and will know that something is wrong if she or he doesn't call or answer.

Volunteers can help older persons with their shopping and can provide transportation or *escort service* to the doctor, church or synagogue, and other places they need to go. If necessary, volunteers can *deliver meals*. They can also do home chores and repairs.

FOOD PROGRAMS

In addition to meals at senior centers and home-delivered meals, neighborhoods run other kinds of food programs (People Power, 1979:21-94; CAI, 1979c). Some are large economic enterprises, such as coop supermarkets, but others fall into the self-help category.

Food buying clubs are found in many neighborhoods. Members get together, decide what they want to buy, and send one member to the wholesale market to buy in quantity. They divide the purchase and split the cost. This method is particularly applicable for buying fresh produce.

A few neighborhoods have organized this the other way around by sponsoring a *farmers' market*, bringing the producers right into the neighborhood on a regular basis. This reduces costs because wholesalers and distributers aren't involved. For reasons of scale, it takes a very large neighborhood or several neighborhoods together to provide a sufficient number of customers for a farmers' market.

Food buying clubs and neighborhood meals programs can benefit from ties to a central warehouse which buys directly from farmers. A considerable number of cities now have central food banks or depositories which receive donations from supermarket chains, wholesalers, and food producers of goods and produce they can't sell, and the food bank makes this food available to nonprofit organizations, many of them neighborhood-based.

Community gardens offer residents an opportunity to grow their own food. They can be divided into individual plots or operate as a communal venture to produce food for a meals program. A few neighborhoods have experimented with greenhouses to grow food year round.

Communal Events

Neighborhood residents also get together to have fun and fellowship, participate in the performing arts, and preserve their cultural heritage.

BLOCK PARTIES

In one sense the historic, congested, immigrant neighborhood was a constant block party because so much of the life in good weather took place on the street. This has lessened but many organized blocks sponsor an annual block party, or several adjacent blocks get together. They get permission to close the street for the day or evening, put up banners, arrange for music (live or recorded), organize games for children, serve food and drink, and enjoy lots of talk and laughter.

NEIGHBORHOOD FAIRS

A block party writ large becomes a neighborhood fair. Usually it is held on playground or schoolyard or in a recreation center or school building. There are likely to be booths for games and many different kinds of foods. Some neighborhoods open their fair with a parade. Live musicians and other entertainers perform. Contests are held for children if there is space. There may be a few speeches and giving recognition awards to key leaders and volunteers. A political figure might be invited to appear. Some neighborhood fairs attract and welcome visitors. In this manner they become fundraising events for the neighborhood association, the local public school, or some other organization (National Trust, 1980; HUD, Public Affairs n.d.).

HISTORY AND CULTURAL HERITAGE

An increasing number of neighborhoods are writing their history. This occurs both in neighborhoods with a particular ethnic identity and in neighborhoods which may have had shifting population or considerable heterogeniety. Long time residents are interviewed. The public library and old newspapers are searched for early history. Old photographs are sought. The product is a book, report, or exhibit which helps give residents a sense of the historical development of their neighborhood, even if it isn't unique enough to be placed on a historic register.

Ethnic groups based in particular neighborhoods likewise research the history of their ancestors. At neighborhood fairs and on other occasions they cook and serve ethnic food, perform characteristic dances, and wear ceremonial clothes from their country of origin.

Some neighborhoods also use contemporary drama and dance as a cultural expression of where they are today.

Analysis

We could go on almost endlessly cataloguing neighborhood activities, but we have reviewed enough of them to get a feel for their variety. Now we can briefly analyze how they originated, what they are, what they accomplish, and where they might be headed.

In 1973 the Washington Consulting Group undertook a federally funded research project for the United States Jaycees Foundation to examine successful self-help projects. Their analysis revealed that successful self-help projects tend to have one or more of the following features (Uplift, 1976:xvi):

- The enterprise is self-sustaining and is often a spur to the development of community programs.
- There are tangible benefits to the community (i.e., improved quality of life, more social services, increased employment opportunities, increased income, better health facilities, better housing).
- The ratio of private to public monies in high.
- The project was community-initiated and remains community-controlled.
- There is significant cooperation and integration with other community organizations.
- There is effective leadership, effective use of volunteers, and sound fiscal management.

In 1979 the Civic Action Institute surveyed 18 neighborhood self-help programs in a variety of program fields. In all of them action was initiated by a few highly committed people, who got things underway and then expanded the participation. It was often necessary to combine different capabilities, such as community leadership ability, technical knowledge, and political connections outside the neighborhood, talents rarely possessed by a single individual. Most of the activities started small and grew to larger programs. Funding was modest in the beginning, but after a while they were able to build a broader base of support. Volunteers were important from the beginning and continued to be significant (CAI, 1980).

Robert L. Woodson found a similar evolution in neighborhood youth programs he studied. He reported (1982b:417):

> In contrast to professional youth projects, urban neighborhood programs are conceived, organized, and maintained voluntarily by community people who are neither connected with the welfare bureaucracy nor trained professionally. In the beginning of the projects, basic economic support comes from the personal household budget of the leaders and is supplemented by neighborhood fund-raising efforts. As the projects mature, some active members subsequently go on to qualify as professionals, and public funds are sometimes obtained for established programs.

That was true of all the programs he surveyed.

As some of these self-help projects evolve and their operations enlarge, their nature changes. They become neighborhood corporations with sizable operations instead of the self-help association of their origin. We will see a number of them in the next four chapters. But many neighborhood projects are content to stay small and remain as largely voluntary activities.

Exercises

(1) Find a neighborhood self-help project now underway. Describe who is participating (volunteers, staff), what they are doing, the budget (if any), whose auspices. Is it growing or likely to remain small? Why?

(2) Interview an initiator of neighborhood self-help. Why did she or he get started? How has she or he gone about enlisting support from neighbors? What problems have been encountered? How overcome? What would she or he do differently if starting again?

(3) Get involved in a self-help project. Or start one. Keep a log of what you have done, relationships with residents and local government officials, how problems have been resolved.

Note

1. This chapter draws upon several collections describing neighborhood self-help projects. Identity in text (and full author in bibliography) is as follows: *Baltimore, 1982* (Citizens Planning and Housing Association); *CAI, 1979a, 1980* (Civic Action Institute); *New York, 1978* (Carlson); *People Power, 1979* (U.S. Office of Consumer Affairs); *Sampler, 1979* (U.S. Department of Housing and Urban Development); and *Uplift, 1974* (Washington Consulting Group). Also see RAIN, 1977; California Office of Appropriate Technology, 1981; and Stokes, 1981. Ongoing sources of information include the following newsletters: *Conserve Neighborhoods* (National Trust for Historic Preservation); *Neighborhood Ideas* (Civic Action Institute); *The Neighborhood Works* (Center for Neighborhood Technology); and *Self-Help Reporter* (National Self-Help Clearinghouse).

CHAPTER 12

DELIVERY AND COORDINATION OF MUNICIPAL SERVICES

Many public and private agencies deliver services within neighborhoods. Some are resident controlled, but many more are decentralized operations of agencies serving a wider territory. Because services interrelate, coordination can enhance their effectiveness, and some aspects of coordination are most effective at the delivery level.

As we noted in Chapter 5, treatment of neighborhood as an administrative unit is one manifestation of its existence as a political community. This can occur in one of three modes of decentralization: administrative, political, and contractual (see p. 67 for elaboration). Or, it can represent noncentralization, that is, operations localized from the beginning.

In this chapter, we examine neighborhood delivery and coordination of basic municipal services. We'll usually speak of city government, but similar observations apply to county government rendering services to neighborhoods. Subsequent chapters look at social services, health and mental health programs, public education, housing, physical and economic development.

As an overview, Table 12.1 provides a listing of services which have somewhere been decentralized, as reported by the Urban Consortium — a network drawing together the nation's 28 largest cities and 9 largest urban counties (Bird, 1981). Some cities set up localized citizen advisory committees to go with city-run neighborhood operations. Some delegate operations to private nonprofit operations. Some municipal services are physically dispersed without any citizen advisory committee involved.

Service Delivery

In our society we utilize local government to provide or arrange for many of the services required for urban living: police and fire protection, water supply, street maintenance, collection and disposal of

Table 12.1
Types of Services Which Some Cities Have Decentralized to Neighborhoods

Function	*Options A or B*	*Option A only*	*Option C*
General	Information, referral Consumer education	Budgeting and priority setting	Tax payment office Licenses and permits
Public safety, corrections	Crime and delinquency prevention Correctional services Abuse and neglect counseling Group homes for offenders	Police	Fire service Courts
Public works	Refuse collection Pest, rodent control		
Recreation	Recreation and leisure-time service Cultural arts Senior centers		
Library	Libraries		
Health	Primary health care Home health care Health education		
Social service	Child day care Individual and family counseling Parenting education Homemaker service Client transportation		
Education	Alternative schools Tutoring Adult education	Public schools	
Housing and physical development	Neighborhood planning and development Subsidized housing Housing maintenance Weatherization Emergency fuel Group homes for special populations	Land use planning Zoning	Housing and building inspection
Employment and economic development	Training, job placement Sheltered and household employment Small business development Commercial area development		

SOURCE: Adapted from Bird, 1981:43-44.
NOTE: Option A — City operation with a neighborhood advisory committee.
Option B — Operation delegated to nonprofit organizations.
Option C — City operation without a neighborhood advisory committee.

refuse, sewage removal and treatment, public health measures, building inspection, recreation, social services, public education. Because of their nature, most of these services are distributed throughout the central city and suburbs. Streets, water and sewer mains, fire stations, playgrounds, recreation centers, libraries, schools, and other community buildings are physically dispersed. Personnel staffing and maintaining these facilities are likewise dispersed. So are field personnel for numerous other services, such as police officers, refuse collection crews, street maintenance units, building and housing inspectors, visiting nurses, health officers, and social workers. They may be based either in field offices or at central headquarters, but their daily routine takes them into neighborhoods.

In this sense, many municipal services can be considered neighborhood services. However, city departments differ in how clearly they focus on neighborhoods and relate to residents. Likewise neighborhoods have differing degrees of influence on and control over these services and the personnel rendering them.

NEIGHBORHOOD APPROACHES OF MUNICIPAL AGENCIES

The predominant pattern is for field personnel providing these kinds of neighborhood services to be on the public payroll. In rendering the services they can deal with residents as merely clients or customers without taking note of neighborhoods and neighborhood organizations. However during the last 20 years, as neighborhoods have become more vociferous and public officials have become more aware of neighborhoods as territorial, political, and social entities, a considerable number of departments in many different cities have initiated new ways of working with and relating to neighborhoods. We highlight a sample of approaches.

Community relations, handling complaints. Most public agencies want to keep on good terms with the people they serve and to gain public support for their activities. Accordingly, they produce and distribute flyers and work with the mass media to achieve a positive image. Many agencies have special central units to receive and deal with citizen complaints. Some agencies, such as police, fire, and public works, also have community relations units to keep in touch with civic groups in order to strengthen their relationships, deal with grievances, and build support (ICMA, 1977b; Brown, 1977). Neighborhood organizations have increasingly become the focus of their activities, and community relations officers spend a lot of time at neighborhood meetings.

Educating the public. Departments seek to educate and persuade the public to take actions which help fulfill their mission. Fire departments teach fire safety at schools and neighborhood meetings and promote use of smoke detectors. The Philadelphia Fire Department, for example, has a mobile unit which goes to the scene of fatal residential fires to show other residents how their homes can be safer. Public works or sanitation departments offer advice on correct storage and placement of refuse containers. Police departments conduct crime prevention seminars. Health departments instruct residents on public health matters, and recreation departments distribute physical fitness information. Various departments sponsor essay contests and conduct tours of their facilities.

Task forces and teams. Sometimes a public agency creates a special task force to work with residents on neighborhood problems, such as sanitation, litter, concern for abandoned buildings and vacant lots. These are task-oriented and usually of limited duration.

Some police departments have organized patrol officers into teams assigned permanently to neighborhood-size beats (Bloch and Specht, 1973; Yin et al., 1976). The teams maintain close internal communication among all officers assigned to the area during a 24-hour period, seven days a week, and they promote maximum communication between team members and the community. A study of police teams in seven cities concluded that the most successful ones featured "unity of supervision, lower-level flexibility in policymaking, unified delivery of services, and combined investigative and patrol functions" (Sherman et al., 1973:5). Team policing failed or received only partial success where it encountered middle management resistance, confronted limits in dispatching technology, and lacked clarity of team behavior and roles (1973:107-108).

Special projects. Field personnel also sponsor special projects which draw in neighborhood volunteers. In fact, many of the self-help activities reviewed in the preceding chapter are frequently promoted by public agencies: block watch and neighborhood patrols by police departments, arson watch by fire departments, cleanup drives and beautification campaigns by public works departments, paint-up, fix-up efforts by community development agencies, sports programs and other volunteer-led leisure-time activities by park and recreation departments, voluntary human services by social service departments. Branch libraries sponsor special events (such as exhibits of neighborhood history and local artists), act as neighborhood information centers, and reach out to special populations, such as children, the aging, the handicapped, adults of limited education.

In Washington, D.C., a special, two-year project linked productivity improvement of street cleaning and refuse collection workers in a pilot neighborhood with increased citizen participation. City personnel worked with the advisory neighborhood commission to educate the public at the same time they were changing operating procedures to get more production without adding any new resources. As a result, citizens improved their practices of putting out trash, streets and alleys were cleaner, compliance time on health code violations was cut in half, litter was perceived as less of a problem, and neighborhood pride increased (District of Columbia Government, 1978).

Special facilities. In some instances, departments open special kinds of neighborhood facilities. Thus, police departments set up small, storefront units to disseminate information and foster better community relations. Eugene, Oregon has established several satellite fire stations, each housing one rapid attack vehicle, which is a small, four-wheel-drive truck with emergency tools (axes, sledges, crowbars, salvage equipment, and fire extinguisher), a 250-gallon water tank, and hoses. These units are handled by two persons, and the stations are designed to look like family dwellings with extra large garages (ICMA, 1981).

Hiring. A major accomplishment of the Community Action Program, set up by the Economic Opportunity Act of 1964, was the widespread use of residents of poor neighborhoods as paraprofessionals in many different roles. The Public Service Employment Program of the Comprehensive Employment and Training Act of 1973 created other kinds of jobs which relatively low-skilled residents could fill. Although these sources of funds are more limited now, there continue to be job opportunities for residents to work as aides in their own neighborhoods, in addition to city jobs open to all through normal hiring practices (which might be through civil service or patronage arrangements). Recreation departments particularly have many part-time jobs for residents. Police departments hire community service aides from neighborhoods served, and some departments have police reserves, who serve voluntarily or for a modest stipend, to help out with animal complaints, parking enforcement, pedestrian safety, and property damage reports (ICMA, 1977b).

DELEGATION TO NEIGHBORHOOD ORGANIZATIONS

Although most basic municipal services are handled by public agencies, certain activities are sometimes contracted out to private for-profit companies, such as for refuse collection, street paving, and other construction projects (Hatry and Valente, 1983). Voluntary fire companies operate in some suburban areas, and a few places have contracted fire fighting to private corporations. In the late 1960s cities began to contract with neighborhood organizations for housing development activities, usually financed with federal funds, as we'll explore in Chapter 14. In recent years a few locales have turned to neighborhood organizations to furnish more traditional municipal services.

Contracting. The most common mode is for the city or county to enter into a contract with a neighborhood organization to perform a specific service. Louisville, for instance, has been doing this with housing development services since 1976. In 1980 when a sidewalk construction project in the Butchertown area ran into difficulty, the city asked the Butchertown Neighborhood Government (a nonprofit corporation) to take it over. The two parties negotiated a $150,000 contract and a second one for $73,000 in 1982 (M. Kotler, 1982:49). Baltimore has a contract with the Park Heights Community Corporation for rodent control and boarding up vacant houses. In St. Paul the District 12 Neighborhood Council has contracted with Ramsey County for solid waste control, designed to separate compostable material from garbage and to make the compost available to residents for their gardens (M. Kotler, 1981:6, 11).

In Philadelphia participants in a series of meetings between the managing director and neighborhood leaders identified the following services which neighborhood organizations could handle: vacant lot cleaning and maintenance, cleaning and securing vacant buildings, minor street maintenance, park and recreation area maintenance. The city administration is following through by awarding five to seven contracts on these matters to qualified neighborhood organizations which submit successful bids (M. Kotler, 1982:48).

Coproduction. Baltimore has developed a Sparkle Program in which neighborhood organizations and city personnel work together on particular projects. Under an agreement with the city, the Mayfield Community Organization hired workers to clean and maintain the stream in Herring Run Park. The city furnishes a truck for hauling away loads of trash. In the Poinsor Hills area a neighborhood organization has an agreement to maintain a city-owned totlot, and the city provides weekly trash pick up. The city government is bringing other neighborhoods into "co-production agreements" (M. Kotler, 1982:48).

Kansas City, Missouri has developed a demonstration program to involve neighborhood organizations in clean-up, snow plowing, housing code inspection, park development, and park maintenance. Neighborhood personnel will be allowed to use city trucks for trash pick-up and snow-plowing equipment. Neighborhood inspectors will check for violations of the city's housing and pest codes and seek voluntary compliance. The city will make available materials needed by volunteers working on improvement of neighborhood parks (M. Kotler, 1983:82-84). (Also see Whitaker, 1980; R. Rich, 1981; Brudney and England, 1983.)

Coordination

Although city departments are part of a single governmental unit, they tend to concentrate on their own functions with little or no relation to what other departments are doing. However, many of the needs of citizens and neighborhoods can properly be met only through coordinated efforts. For example, street cleaning requires the traffic department to post signs prohibiting parking during certain hours of certain days, the police department to enforce the prohibition, and the public works or street department to sweep the streets.

Mayors and city managers are aware of the need for coordination and have instituted a number of neighborhood-level methods. Mayors are also interested in maintaining good working relationships with citizens for political reasons.

LITTLE CITY HALLS, MULTISERVICE CENTERS

Some cities have established little city halls (called by various names) as a means of reaching neighborhood residents more effectively, and even more cities have set up multiservice centers aimed at bringing together varied service personnel. This is by no means a new approach, for Boston erected district municipal buildings in the early part of this century and Los Angeles began establishing branch city halls in the 1920s, necessitated by the spread of that sprawling city. However, this practice accelerated in the 1960s, particularly in response to racial unrest and citizen discontent with municipal services. Both the National Advisory Commission on Civil Disorder (1968:294) and the

National Commission on Urban Problems (1968:351) thought this approach to neighborhood decentralization merited attention, and a number of federal agencies gave supportive grants.

A 1971 survey by the Advisory Commission on Intergovernment Relations (ACIR) revealed that little city halls and multiservice centers were more prevalent in the larger cities, as might be expected:

	Percent of cities with:	
Population group	*Little city halls*	*Multiservice centers*
Over 500,000	40	60
250,000 – 500,000	22	72
100,000 – 250,000	6	33
50,000 – 100,000	4	19
25,000 – 50,000	0	8

The five most common functions stationed at little city halls were police (mainly community relations), complaint and information, streets, sanitation, and housing code inspection. The five most common functions handled by multiservice centers were community action, senior citizen activities, employment services, welfare, and health (ACIR, 1972:10-11). There has been no subsequent national survey, but probably the use of little city halls has decreased somewhat and the use of multiservice centers has been stable or grown slightly.

What are called little city halls deal mainly with municipal services while multiservice centers bring together health and welfare services (the subject of the next chapter). Little city halls differ according to relationship with the chief executive (mayor, manager, or chief administrative officer), how many departments locate personnel there, and the relationships among these personnel.

Mayoral unit. Some little city halls have functioned principally as units of the mayor's office. Boston went the furthest along these lines from 1968 until 1981, operating 15 little city halls with 135 people and an annual budget of $1.5 million, until a severe fiscal crisis caused their elimination. Each was headed by a manager, who served as the mayor's representative to facilitate flow of communications between the mayor's office and citizens, and vice versa. The managers ran an information and referral service, handled complaints, and promoted special neighborhood projects. They also helped develop political support for the mayor in reelection campaigns. Residents could go to little city halls to pay sewer, water, and property taxes, get documents notarized, obtain marriage, birth, and death records, apply for city jobs, and register to vote. Some of them provided space to city inspectors, but they did not emphasize colocation of various services. Linkages with operating departments came through the Office of Public Service, a part of the Office of Mayor in city hall. (For early history, see Norlinger, 1972; Washnis, 1972:235-279).

In the late 1960s and early 1970s similar neighborhood-based, mayoral units, though of lesser scale, were tried in other cities, such as New York, Baltimore, Atlanta, Houston, and Columbus, Ohio. However, this mode of decentralization never became widespread, though there are still some places, such as Detroit and Baltimore, where the mayor has liaison staff based in neighborhood offices.

Colocation. Another approach is to use little city halls as bases for field personnel from city departments. (District county buildings serve the same function for county departments.) The personnel may be under the same roof for convenience of location with relatively little interaction among them, or there may be a manager or coordinator who seeks to orchestrate their work.

Los Angeles' 11 branch city halls provide an example of colocation without much coordination. They range in size from the nine-story Van Nuys Municipal Building in the outlying San Fernando Valley to several one- and two-story buildings with one quarter the space. Fifteen different departments assign personnel to one or more branch city halls. County and municipal courts are in two of them. Nine provide offices for council members, the mayor has a representative in two, and one houses offices of state legislators. Several make space available to voluntary organizations.

Coordinated services. Some neighborhood municipality facilities go beyond colocation and seek to coordinate the services provided by the personnel based there. This occurs in Baltimore, a city which started two tracks in the 1960s: mayor's stations to house personnel from a number of municipal departments (Washnis, 1972:281-302), and multipurpose centers with more of a social service orientation, financed by the Community Action and Model Cities Programs. The latter also drew federal construction funds from a community facilities program administered by the U.S. Department of Housing and Urban Development. By now the two thrusts have merged and mayoral appointees serve as managers of Baltimore's 15 multipurpose centers and coordinate the work of various municipal and social services.

ORGANIZATIONAL ARRANGEMENTS

Some cities try to coordinate neighborhood services through organizational arrangements not depending upon colocation.

Task forces. One method is to appoint a task force of public officials, and possibly citizens and representatives from private organizations as well, to address particular problems in specific neighborhoods. New York pursued this approach from 1968 to 1970 through a set of urban action task forces, each assigned to a community district (Washnis, 1972:158-166). They consisted of field personnel from city departments and citizen leaders, and each was chaired by a top official from the mayor's staff or a city department. Exposure to community concerns gave city officials a better view of neighborhood problems, and the rank of the chairpersons provided some clout in developing solutions. They succeeded in taking care of a variety of problems, but the structure was too unwieldy for long-term operations. Experience elsewhere indicates that to succeed task forces require strong leadership, a tight focus, deadlines for action, and collective influence.

Minicabinets. New York then embarked upon an experiment using district service cabinets composed of field supervisors of the departments of police, sanitation, parks, recreation, highways, health, and human resources. A district manager, appointed by the mayor, chaired the cabinets and was responsible for following through on matters requiring joint action. In this same period (the early 1970s) the District of Columbia set up area services committees consisting of representatives of similar departments (Arnando and Peel, 1974), and Dayton established administrative councils to work in areas served by neighborhood priority boards. In the latter two cities participants included persons based in city hall as well as field personnel.

An evaluation of the New York experience indicated that a preponderant majority of field personnel felt that their monthly cabinet meetings were useful. Many projects they undertook succeeded in improving service delivery. Three out of four experimental districts were more successful than the fourth, and they gained a favorably community response. The greatest difficulties stemmed from insufficient delegation of authority to field personnel (Barton et al., 1977; Katznelson, 1981).

District managers. The presence of a district manager was an important factor in the New York experiment. This is a person who is not tied to any single department and has connections with the mayor or manager. This enables him or her to cut across departmental lines and to have ways of influencing top-level departmental decisions. This role can be fulfilled either by a manager of a multiservice center or a person presiding over a district or neighborhood cabinet or task force. Interagency coordination is a difficult task and depends to a considerable extent upon persuasion and willingness of participants to work together rather than authoritative command. However, it helps to have strong backing from the chief executive, positive support from department heads, and outside pressure from citizens.

COTERMINOUS DISTRICTS

One obstacle to neighborhood coordination is the lack of similar boundaries for the administrative districts of different departments. Ray N. Bird has observed that departments use one or more of the following criteria to draw districts (1981:9):

- The number of area superintendents the agency has the money to hire and/or can adequately supervise in standard organizational "span of control" practice.
- Minimum number of staff and cases required to organizationally justify a district office area.
- Areas of the city that can produce effective political pressures on an agency for an area office in their neighborhood.
- Externally imposed criteria, such as federal program standards.

Because departments develop their individual rationale, a city map of all district boundaries presents a bewildering pattern.

Washington, D.C., has dealt with this problem by using its eight wards as service area boundaries for most departments. This provides approximately equal population size but in some instances cuts across natural neighborhood lines. New York City has 59 community districts with a population range of 50,000 to 260,000, with boundaries determined through negotiations of diverse interests. Major departments have realigned their service boundaries accordingly.

Citizen Involvement

PUBLIC AGENCY PERSPECTIVE

As part of their community relations activities, neighborhood administrators sponsor neighborhood meetings and attend meetings called by neighborhood organizations. They form tasks forces and conduct workshops to deal with specific issues. Agencies with ongoing services may set up neighborhood advisory committees, a practice followed by some (though not all) police districts, recreation centers, health centers, and community development project offices.

The ACIR survey, cited earlier, found that 70 percent of multi-service centers had resident advisory boards while only 29 percent of little city halls did (1972:12). Washnis' study noted that the latter were more likely to be the mayor's representative in the neighborhood, and little city hall managers preferred to work through direct personal contact with residents and neighborhood organizations rather than set up formal advisory committees (1972:363-366).

RESIDENTS' PERSPECTIVE

Residents insert themselves into decentralized administration by using three modes of influence (see p.144): as interest groups, through the electoral process, and by getting involved in structured citizen participation.

Each neighborhood has a variety of interest groups. Some are geographic, such as block groups and neighborhood associations. Some revolve around the manner of tenancy: homeowners, tenants, condo owners, coop members. Some have a problem or program focus, such as crime prevention, education, housing conditions, economic opportunity. As advocates of particular interests, these groups seek out neighborhood-based administrators, communicate with them, and try to influence their decisions. They might also deal with their bosses at central headquarters in order to affect the course of decentralized administration.

They also work through political channels. Where precinct and ward leaders are influential, residents turn to them. Or, they go to a member of city council or the mayor with their problem. And neighborhood organizations invite elected officials and candidates to their meetings in order to get across their views.

Residents show up at neighborhood hearings, urge field offices to set up advisory committees, get involved in neighborhood-based planning activities, and take delegations to city hall hearings.

NEIGHBORHOOD COUNCILS

Cities with officially recognized neighborhood councils provide specific roles for them in connection with decentralized administration. Thus, in New York each of the 59 community districts has an appointed community board which selects a district manager, who serves as chairperson of a district service cabinet consisting of field supervisors from the departments observing community district boundaries for their field operations. In Dayton the Division of Neighborhood Affairs has a site office in each neighborhood priority board area, works with and provides staff services to these boards, and convenes the administrative council of departmental representatives each month. In Birmingham representatives from the police department and other agencies meet regularly with the elected neighborhood advisory committees. Routinely in most cities with neighborhood councils, city departments refer for comment all proposals for new projects and facilities in the neighborhood. (More on neighborhood councils in Chapter 16.)

Analysis

When Robert K. Yin and Douglas Yates reviewed 215 case studies of decentralization in five urban service areas (public health, safety, multiservice programs, education, and economic development), they concluded that the cases showed the following results (1974:59-68):

Outcome	*Percentage of cases*
Improved services	72.1
Increased information	63.2
Improved client attitude	32.1
Increased client control	24.9
Improved services officials' attitudes	17.3

Their analysis divided the cases into weak, moderate, and strong forms of decentralization. The weak forms (30.7 percent of the cases) included physical deployment and administrative decentralization without client participation, moderate forms (26.0 percent) encompassed indigenous employment and new neighborhood institutions, and strong forms (43.3 percent) occurred where clients had direct governing control over a service delivered to a specific neighborhood. Improved services occurred under all three forms but happened with the greatest frequency with the strong forms. The weak forms placed the greatest emphasis upon increased information (1974:64-82).

Other studies have noted that neighborhood involvement has contributed to improved service delivery, particularly for sharply focused programs (Yates, 1973; R. Cole, 1974). Also, administrative decentralization with citizen participation has contributed to leadership development, producing new skills leading to employment opportunities, and achieving a greater sense of personal dignity, self-worth, trust, and political efficacy on the part of participants (National Commission on Neighborhoods, 1979: App. 1; R. Cole, 1981). (We'll look more fully at these findings in Chapter 16).

In his review of municipal services contracted to neighborhood organizations, Milton Kotler concluded that these kinds of arrangements may be less costly to local government than direct municipal operation, permit tailoring of services to specific community needs, and make possible the integration of volunteers and part-time workers. They offer neighborhoods a way of protecting service levels in a period of retrenchment, a source of income for their organization, jobs for residents, and the possibility of new enterprises as offshoots as they seek broader markets for similar services (such as refuse removal from businesses and industries, private landscaping, energy audits for private companies). However, there are risks of poor neighborhood management, political awards to unqualified organizations, and service interruption if the contractor changes. Impediments to broader use of neighborhood contracting include no-layoff contracts with public employees, legal barriers to sole source contracting, problems of who is liable for injuries to persons and property, administrative difficulties in setting up a new system, and political opposition from employee unions (M. Kotler, 1983:14-38).

Apart from reluctance of municipal agencies to contract services to private entities, many neighborhood organizations don't particularly desire taking on responsibility for service delivery. Compared to the late 1960s, neighborhood activists are now less inclined to demand control of municipal services. They prefer to act as advocates and put pressure on the departments for high quality service without getting bogged down in administrative responsibilities themselves. That is to say, they favor an influence strategy rather than gaining direct control. This orientation has combined with city resistance so that contracting municipal service delivery to neighborhood organizations is only in the experimental stage in a few cities and not a widespread pattern.

A city government also has tradeoffs to consider in moving towards greater neighborhood decentralization. For instance, the initial New York experiment with district managers used them as representatives of the mayor, charged with achieving district-level service coordination. The community boards were fairly passive participants in this process, instead concentrating more on city planning issues. Then a charter amendment strengthened the powers of community boards, particularly enhancing their role in the budgetary process and giving them authority to appoint the district manager. Every community district got a manager, but their salaries were lower than those in the experiment, in fact below that of departmental field supervisors, and they no longer had their direct connection with the mayor's office. This lessened their authority over field operations and reduced the effectiveness of district cabinets in service coordination. However, the change has provided greater citizen access to district operations on particular complaints (Mudd, forthcoming).

Where neighborhood service operations are contracted to neighborhood organizations, greater administrative dispersion occurs and may make service coordination more difficult. However, coordination through mutual consent and contractual obligations is still possible even though the clout of line authority is reduced.

Neighborhood decentralization occurs for a multiple of reasons and requires compromises among diverse objectives and values.

Exercises

(1) Identify a neighborhood service run by a public agency without much citizen involvement and another which has a larger role for citizens.

(2) Describe the nature of the service, who is in charge, what kind of personnel are involved, where they are based, the extent of administrative discretion they possess.

(3) Describe the way citizens are involved, which citizens, what kind of structures, and the relationships between citizens and service delivery personnel, and the extent of citizen influence.

(4) Assess the effectiveness of the services you are studying and the difference citizen participation seems to make (if any).

(5) Find an example of neighborhood services coordination. Describe the services being coordinated, the participants, who is in charge, the methods of coordination, what coordination has achieved, its shortcomings.

(6) To what extent is coordination based upon line authority and clout, persuasion and shared objectives, or mutual self-interest?

CHAPTER 13
HUMAN SERVICES AND EDUCATION

Another set of services available to urban residents are those which help individuals and families to solve specific problems affecting their physical, mental, and social well-being. This chapter examines how such human services are delivered and coordinated within neighborhoods. They cover the entire span of life, serving all ages and many different subgroups of the population, as shown in Table 13.1. The list is long but not necessarily exhaustive, for it doesn't detail numerous service specialities.

All these human services aren't available in every neighborhood, some are found both in neighborhoods and outside, and specialized aspects may be available only in centralized facilities. Many of the services are part of broader systems. They are handled by both public and private agencies, including several layers of government; neighborhood based, citywide and metropolitan nonprofit organizations; individual proprietors, partnerships, and for-profit corporations; and cooperatives. In this chapter we look at four major clusters of human services: social services, health, mental health, and public education.

Social Services

In Chapter 8 we touched briefly on the development of social services in response to problems and needs caused by industrialization and urbanization. Some new services were neighborhood-based, though often initiated by a citywide organization. For many years private nonprofit organizations led the way, but since the 1930s public agencies have become a larger influence. There has been a long term movement toward professionalism and specialization. By the 1950s these trends resulted in minor roles in social service delivery for indigenous, neighborhood organizations, and among the established, professional agencies, only settlement houses were consistently neighborhood-based.

When the Economic Opportunity Act of 1964 initiated the War on Poverty, the U.S. Office of Economic Opportunity (OEO) offered the following critique of the service delivery pattern in poor neighborhoods, and proposed a set of responses revolving around neighborhood centers (1966:2-10):

(1) Problem: People don't know about services.
Answer: Outreach, referral, and follow-up.

(2) Problem: Services are far away.
Answer: Decentralization of services and programs.

(3) Problem: Services and programs are fragmented.
Answer: Coordination of services.

(4) Problem: Existing services are inadequate.
Answer: Modification and improvement of services.

(5) Problem: The poor are treated as second-class citizens.
Answer: Neighborhood involvement.

OEO had funds in the Community Action Program to initiate responses, a mandate to involve residents from areas served, and for

Table 13.1
Human Services Suitable for Neighborhood Delivery

Family planning
Prebirth counseling for parents-to-be
Parenting education
Child day-care
Preschool programs
Child protective services
Services to children with special needs (physical disabilities, speech and learning problems)
Elementary and secondary education
Recreation for all ages
Adolescent counseling
Services to juvenile delinquents
Counseling for adults and families
Crisis intervention
Parole and probation services
Home services for the aging and chronically ill
Respite care
Bereavement counseling
Services to adults with physical disabilities
Adult education
Vocational rehabilitation
Employment training
Job placement
Income maintenance
Emergency funds, food, shelter
Nutrition services
Health education
Health screening and diagnosis
Primary health care
Home health care
Hospices
Mental health diagnosis, counseling and treatment
Substance abuse programs
Group homes (halfway houses) for special populations, such as mentally ill, juvenile delinquents, released offenders, participants in drug abuse programs
Legal services

several years zeal for experimentation. There were many other like-minded actors around the country. The result has been a transformation of neighborhood social service delivery during the past 20 years.

DECENTRALIZED DELIVERY

There has been a substantial increase in the number and variety of social services available in poor neighborhoods. Federal funding has been an important factor, until the cutbacks initiated by the Reagan administration, but local agencies supported by the United Way and foundations have also directed more resources to neighborhood-based operations and some neighborhoods have raised their own funds and mobilized volunteers.

Neighborhood location. Public agencies, such as the state or county welfare department, have opened neighborhood offices, and so have large voluntary agencies, such as family service and child welfare agencies. New neighborhood-controlled organizations have emerged. The facilities vary, such as storefronts, single-service offices or a cluster of related services, and large multiservice centers. Sometimes neighborhood social services operate as an ancillary to another program, such as in a school, employment center, or health center (Vigilante, 1976).

Use of paraprofessionals. Neighborhood social service providers have hired residents to fill positions as aides, or paraprofessionals. As Clifford Shaw discovered in the 1930s (see p. 113), indigenous workers understand the neighborhood culture, have natural contacts with fellow residents, and an intuitive understanding of their needs. Thus, they can undertake outreach, handle initial interviewing at intake, provide supportive services as part of follow through, offer home services (such as home health care and homemaking services to the aging and chronically ill), work with groups (such as youth gangs), and take on other subprofessional tasks. Some agencies have built career ladders for paraprofessionals, and some universities have programs enabling them to obtain "new career" degrees and move into the ranks of professionals.

Resident involvement. Most of these neighborhood programs provide a role for residents in policymaking or review. This ranges from an advisory committee with little power and not much influence to a board of directors in control of a neighborhood agency. In some places representatives of poor neighborhoods have been appointed to citywide boards.

Scale and specialization. Out of this experience has come a realization that many, perhaps most, social services can operate at a neighborhood scale. That is because the unit served is one person or one family and many of the services by their nature require individualization. However, there are some specialized services needed by relatively few persons which require a larger base to be economical and have the necessary skilled personnel. And many neighborhood services must be tied to central operations, such as probation services related to the courts. Considerations of equity and economic redistribution require that income maintenance through welfare, social security, and unemployment compensation be financed by a broad base, indeed national, though intake can occur at a neighborhood office.

COORDINATION

The proliferation of specialized social services, citywide and within neighborhoods, has raised the need for coordination. That's because some personal and family problems are multifaceted and require solutions involving a number of service providers. Several methods of coordination are available (Agranoff, 1977).

General counselor. One approach is to have a general counselor serve as the primary contact with an individual or family and to orchestrate the other services. This approach has been used by caseworkers involved in multiproblem family projects, which have focused upon families with a range of needs demanding services from many different agencies. To be successful, the other agencies must be willing to accept the caseworkers as the primary contact, provide timely service, share information, and if necessary participate in case conferences.

Multiservice centers. Another way is to establish a multiservice center housing a wide variety of services. OEO and the Model Cities Program favored this approach and provided operating funds for numerous installations. Some state, county, and city welfare departments have also sponsored such facilities (Marr, 1973).

According to Daniel Thursz and Joseph L. Vigilante, neighborhood service centers have two major functions — access and direct services (1978:14):

> The former provides the means to link the client with the array of services, and the latter provides the specialized service required. Access services include, for example, information, advice, referral, community and group education, follow-up and escort service, case advocacy, policy

advocacy, institutional and individual linkage, and social brokerage. Direct services include the provision of housing, employment, day-care, legal services, child placement services, personal and family counseling, and other services.

Multiservice centers operate in various ways. Some of them are like shopping malls with many individual operating units located in the same building with relatively little interaction among them. There may be a central receptionist to offer direction, but otherwise service recipients have to find their own way to the services they need. Other multiservice centers have a central intake unit where interviewers make a preliminary diagnosis and refer the recipients to appropriate services. The center may have a manager who deals mainly with building operations, or it may have a director whose job is to coordinate service delivery. Rarely are all employees on a single payroll so that coordination depends upon agreed processes rather than line command. Unit supervisors might confer as a service center cabinet, and counselors get together in case conferences.

Networks. Where all neighborhood services aren't located in a single building, coordination can be achieved through a network approach. As indicated by the National Commission on Neighborhoods (1979:237-238):

> In this model, public and private agencies provide services to residents at a number of locations inside and outside of the neighborhood, with linkages maintained primarily through information, referral, and transportation services. A typical services network in a given neighborhood would include a local community center which may have operated formerly as a settlement house, a mini-city hall which houses public health and welfare agencies, and a community hot line organized and operated by volunteers in an ecumenical ministry center. Other linkage points for special target groups such as senior citizens, handicapped people, and ex-offenders would be provided through local centers designed to service these groups. . . .
>
> The network model provides a process for linking a broad range of public and private agencies that receive funding from a variety of sources. Neighborhood involvement in the operation of the network would take the form of citizen participation in planning, monitoring, and evaluating individual programs and the total system. This process could help determine which services overlap or duplicate one another and where gaps need to be filled with new or expanded services.

It is also possible to develop smaller networks around clusters of services aimed at particular populations, such as services for youth, families, or the aging. A related emphasis is linkage between agencies and families and other primary groups (Litwak, 1978).

Health Centers

Neighborhood residents receive health care from a complex system with numerous components, ranging from private physicians in solo practice to huge medical school hospitals, and financed by payments from individuals, insurance companies, government revenues, and foundation grants. Many components of the system are located outside residential areas, but some physicians, dentists, visiting nurses, clinics, and nursing homes are neighborhood-based. In this chapter we are particularly interested in where a cluster of services come together in neighborhood health centers, usually in poorer neighborhoods.

TWO WAVES

Neighborhood-based health services aren't new. For instance, in 1893 Hull House in Chicago started a public dispensary, the Nurses' Settlement opened on Henry Street in New York as a base for public health nurses in the Lower East Side, and also in New York milk stations began operating because of a concern for the health of immigrant children. In the next 15 years some milk stations evolved into child health centers and set the stage for more comprehensive neighborhood health centers. The first one opened in Milwaukee in 1911, and by 1920 there were 72 centers in 49 cities, mostly in immigrant neighborhoods. The continued to spread during the 1920s. Thereafter, though, they declined as second generation residents turned more to private physicians, the Great Depression curtailed local funding, the New Deal programs took a different emphasis, and health insurance plans came into being (Rosen, 1978; also see Stoeckle and Candib, 1974).

After World War II, federal-aid programs expanded, private health insurance became more widespread, and expansion of medical schools added to the supply of doctors, but some of the older neighborhood health centers continued to function and places such as Philadelphia built some new ones. Nevertheless, poor people still weren't receiving sufficient health care, and a new wave of health centers came forth in the 1960s. The U.S. Office of Economic Opportunity used research and demonstration money to fund eight centers in 1965, and the following year Congress amended the Economic Opportunity Act to authorize comprehensive health projects as a regular component of the Community Action Program. The Partnership for Health Act of 1967 created a similar program within the Department of Health, Education and Welfare. In 1974 HEW took over the OEO program, and by then there were approximately 100 federally funded neighborhood health centers in operation (Hollister, 1974: 1-12). The program got a further boost in 1976 when Congress enacted a new urban and rural community health initiatives program which funded 872 local projects in the next four years. From this zenith the program has been cut back by the Reagan administration. But nevertheless in the spring of 1983 there were 524 federally assisted community health center grantees (177 urban and 347 rural) with operations at more than 800 sites (for some of the grantees had centers in several locations). They were serving nearly 5 million people. Also, some other neighborhood health centers operate without federal aid.

CHARACTERISTICS

The original OEO model of a neighborhood health center had the following characteristics (Schor and English, 1974:46-47).

(1) Focus on the needs of the poor.
(2) A one-door facility, readily accessible in terms of time and place, in which virtually all ambulatory health services are available.
(3) Intensive participation by and involvement of the population to be served both in policymaking and as employees.
(4) Full integration of and with existing sources of services and funds.
(5) Assurance of personalized, high-quality care, and professional staff of the highest caliber.
(6) Close coordination with other community resources.
(7) Sponsorship by a wide variety of public and private auspices. Among the sponsors were community action agencies and medical schools, usually with governance by a board of directors bringing together residents and health experts from the broader community.

This basic organizational pattern has now been incorporated into federal law. To receive federal assistance, a community health center must be governed by a board with a majority who are nonproviders of health care services and who are also users or potential users of the center's services. They must provide 15 basic primary care services. Until funding cutbacks after 1980, they could also use federal money for supplemental services, such as hospital inpatient care and environmental health screening. A typical urban health center has a budget of $3 million a year, though some are considerably larger, some smaller. About 40 percent of their budget comes from federal grants with the balance paid from local sources, user fees, and reimbursement under federal assistance programs to individuals.

AN EXAMPLE

Although neighborhood health centers vary in what they offer, the health services delivered by the Dorchester House/Multi-Service Center in Boston provides an illustration (Jacobson, 1983:11):

Preventive and educational

Consumer health education
Health screening and diagnosis
Physical fitness programs
Self-help peer groups

Primary care delivery

Emergency medical services
Primary health care
Out-patient health care
Out-patient mental health treatment

Auxiliary care

Respite care
Terminal or hospice care

Other operations at the center provide a variety of social services.

The board of governors of the Dorchester House Multi-Service Center consists of representatives from six committees (health, mental health, special needs, preschool and day-care, recreation, and arts) and three at large. Neighborhood residents are eligible to serve on the six program committees and may be elected to the board at a general meeting each May. The board also has four nonvoting members: a banker, a local elected official, a representative of the city's Department of Health and Hospitals, and a university professor (Jacobson, 1983:45-46).

The board of governors has broad powers over the operation of the center, including preparing the operating budget, hiring and firing full-time staff, and determining what programs to provide. The health center uses staff who are on three different payrolls: Federated Neighborhood Houses of Dorchester, Boston Department of Health and Hospitals, and the state mental health agency. They all work under the direction of the center's director. The board has a partnership agreement with the Department of Health and Hospitals committing the Department to provide medical support services (including services and staff privileges at Boston City Hospital), monitor and evaluate the center according to jointly set standards, and furnish technical assistance as needed (Jacobson, 1983:47-49).

Mental Health

A third body of experience in neighborhood human services has developed in the mental health field.

COMMUNITY MENTAL HEALTH CENTERS

A major stimulus was the Community Mental Health Centers Act of 1963, which provided a new source of federal funds for construction and the eight initial years of operation. The Act grew out of a realization that large state mental hospitals weren't the best facilities to treat many mental illnesses, much less do anything about prevention. Community-based care and treatment was seen as a visible alternative.

The 1963 Act envisioned 2,000 community mental health centers in the United States, serving catchment areas of 75,000 to 200,000 persons (the term "catchment" derives from the interest of water resources management in drainage areas flowing into reservoirs). Though later the goal was scaled back to 1,500, only 768 had been established by 1982 when the program was folded into a broader state block grant program and lost its strong federal mandate.

In 1978 President Carter's Commission on Mental Health reviewed the community mental health center program and concluded that the local centers had made significant contributions in their communities. However, it pointed to federal inflexibility in requiring a specific set of services not necessarily needed in every locality and recommended greater local choice. The commission observed that arbitrary population size of the catchment areas fail to take into account natural communities and create unnecessary barriers for both those who need and those who provide services. And it urged greater emphasis upon utilizing "personal and social networks of families, neighbors, and community organizations to which people naturally turn as they cope with their problems" (President's Commission on Mental Health, 1978:14, 17, 62).

Dissatisfaction with some aspects of community mental health centers has led to alternatives more firmly rooted in neighborhood life. For instance, Herzl R. Spiro (1969) proposed a three-level model: (1) comprehensive neighborhood health centers, or satellite mental health clinics, to serve populations of 15,000 to 30,000 and provide intake, outpatient care, and home visiting; (2) community mental health centers to provide inpatient services and intensive rehabilitation for areas of 70,000 to 200,000 people (the federal catchment areas); and (3) a specialized wing of a community mental health center, a state hospital, or a nursing home system to furnish specialized inpatient care for such special categories as drug abusers and the mentally retarded, to serve an area encompassing up to one million people.

NEIGHBORHOOD HELPING NETWORKS

As a further alternative, Spiro teamed with Arthur J. Naparstek and David E. Biegel in a five-year, two-city demonstration program which developed a neighborhood mental health empowerment model not dependent upon a building facility (Naparstak et al., 1982). Their approach was founded on an assumption that every neighborhood, no matter how heavy its problems, has strengths and a natural helping network of people who help individuals to deal with their personal problems. This indigenous network needs to be linked to professionals and to service agencies in a support system operating as a partnership, based upon mutual respect of all parties for one another. Because of ethnic, class, and racial differences in pluralistic America, community mental health systems should take various configurations, tailored to the neighborhoods they are serving. Figure 13.1 shows this system schematically.

One part of the demonstration took place in South East Baltimore, a diverse ethnic community of people from southern and eastern European origins. The other part was carried out in Southside Milwaukee where Polish-Americans are the dominant ethnic group, though also with a German-American population. In Baltimore the South East Community Organization (SECO), a large, diversified body, sponsored the project and formed a neighborhood and family services task force as the community's vehicle for participation. Southside

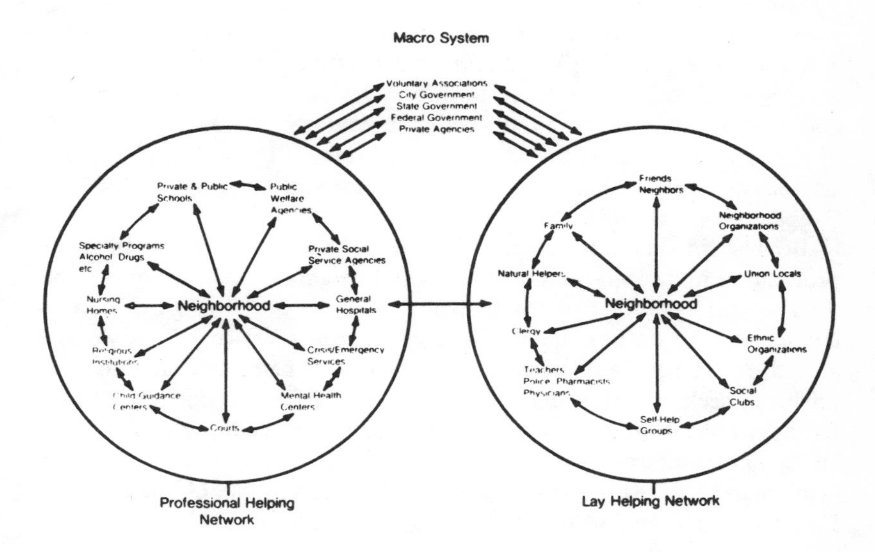

Figure 13.1 Neighborhood Mental Health Empowerment Model

Milwaukee lacked a similar umbrella group, so a local organizer had to put together a coalition, which named itself the South Community Organization (SCO). Both communities formed parallel professional advisory committees, consisting of mental health and human services personnel serving the neighborhoods. Although the lay and professional task forces and committees met separately in the beginning, as the demonstration progressed they joined forces.

The activities which emerged focused on individuals, families, and neighborhood networks. In Baltimore projects included a neighborhood and family day picnic, a hotline, referral directory, clergy-agency-community seminars, case study/brown-bag luncheons, council of human service providers, a peace at sundown program, a babysitting cooperative for young mothers, bus trips for the elderly and shut-ins, and a series of workshops on stress. In Milwaukee SCO gave priority to developing a referral directory to inform residents of existing services. Other projects were a family communications workshop; support groups for divorced, widowed, and agoraphobics; a "wellness" emphasis; clergy/agency luncheons; and a community helpers newsletter (Naparstek et al., 1982:133, 150).

Because in our society mental illness carries a stigma, both local programs emphasized human services and neighborhood support rather than mental health per se. But by being available, mobilizing forces, and connecting the natural helping and professional networks, they were able to reach people in need of assistance who had been unwilling or unable to seek professional help. They helped residents cope with stress, family problems, and other pressures on their mental and physical well-being. (For more on support systems and self-help, see Caplan, 1974; Caplan and Killilea, 1946; Gartner and Reissman, 1980; Gottlieb, 1981; Froland and others, 1981; Schulberg and Killilea, 1982.)

Public Education

Schools are an important part of neighborhood life, especially elementary schools, both public and parochial. Secondary schools are more likely to serve community districts of several neighborhoods and have less of a neighborhood connection. Although some suburban school districts contain a single high school, most urban systems are larger. Our attention here is focused mostly on public schools operating in a neighborhood setting, but we must give some heed to overall organizational patterns of school systems.

SYSTEMWIDE

The majority of school systems in the United States are organized independently of municipal government and have elected school boards, though in some larger cities the system has a structural relationship with city government and the board may be appointed. Neighborhood residents and other citizens have available the three modes of influence outlined in Chapter 10: electoral, as interest groups, and through structured participation.

After studying patterns of participation in Boston, Atlanta, and Los Angeles, Marilyn Gittel concluded (1980:241):

> Citizens have little influence on the educational decision-making process. In lower-income communities there is a lack of political action-directed organizations, coupled with frustration over or disinterest in school issues. The environment of city political life and the operating styles of the school systems do not support or encourage effective citizen participation in major school policy issues. Citizen access is generally limited; it is especially closed to those who actively seek a redistribution of resources throughout the system.

In smaller cities and the suburbs citizens seem to have a greater impact on school decisions, but the bureaucratic nature of school organization and the professionalism of educators make systemwide influence difficult to achieve (Summerfield, 1971).

AREAS OR SUBDISTRICTS

Because of frustrations with centralized school systems, minority groups and other citizen activists in the 1960s advocated greater community control of schools (Levin, 1970; Fantini et al., 1970). For other reasons, larger school systems themselves have moved toward administrative decentralization internally (LaNoue and Smith, 1973).

Community control. New York in 1969 and Detroit in 1973 reorganized into two-tiered systems through action of the state legislatures. Control over policy, personnel, finance, curriculum, and administration was divided between the two tiers. This gave New York 32 community districts, each with an elected school board to oversee elementary and junior high schools, and a central board appointed by the mayor and five borough presidents. The Detroit school system was divided into eight regions, each with an elected five-member board. Each regional board was represented on the central board along with five members elected citywide. However, in January 1983 these regional boards were abolished, and the system was recentralized.

No clear-cut evaluation of the impact of reorganization on student achievement is available for these two cities, but the change has opened more administrative positions for blacks and Hispanics (groups which constitute the majority of school enrollment), has produced more community interest in education, and has brought about some curriculum changes oriented toward needs of the community districts. The change also mitigated severe political conflict between community activists and school officials (Glass and Sanders, 1978; Seeley and Schwartz, 1981; Rogers, 1982).

Administrative decentralization. A more common pattern for the bigger school systems is to institute administrative decentralization by appointing superintendents over areas or subdistricts within the larger system. A survey of 62 large systems in the early 1970s found that in addition to New York and Detroit, 27 out of 35 school districts with enrollment over 75,000 had decentralized or were considering it, and that 8 out of 25 districts in the 50,000 to 75,000 enrollment range were likewise involved with administrative decentralization (Ornstein, 1974). Basically, this has occurred internally within an administrative structure which is hierarchical in nature, shifting some authority downward to area superintendents. Rarely has citizen participation increased, for the area superintendents usually don't have advisory committees, and they are probably less known and accessible to citizens than central personnel.

INDIVIDUAL SCHOOLS

Parental and citizen influence is likely to be the greatest at the neighborhood level in relation to individual schools. It may also be the place where administrative decentralization is more important.

School-based management. On the latter issue, some school systems, such as in Florida, California, and Kentucky, have moved toward school-based management, defined by Don Davies as (1981:104):

> making the individual school site the cost center by means of lump-sum budgeting to the site, requiring cost accounting and school performance reports at each site, and increasing principal accountability for individual school program budgets. . . . It may include having building principals responsible for hiring, assigning, and firing school employees at the building level and allow for school-by-school curriculum planning.

This kind of decentralization is ordinarily accompanied by a greater role for parents, other residents, and teachers in school policymaking.

School advisory councils. Most schools gain some parental participation through parent-teacher associations or home-and-school councils, and many bring them in as volunteers, especially in middle class neighborhoods. Some federal-aid programs have mandated advisory councils, systemwide or for individual schools. The legislatures of Florida, South Carolina, and California, as part of school finance reform and out of a concern for educational accountability, have mandated school councils at the building level. The councils provide membership for parents, teachers, and, at secondary schools, students. School councils in Florida and South Carolina participate in preparing an annual report of school progress. They function in California in schools participating in a special, state-funded school improvement program and have responsibility for designing improvement plans and overseeing their implementation. A study by the Institute for Responsive Education concluded that these school councils are able to function not merely as advisors or watchdogs but more like active partners with school personnel (IRE, 1980; Zerchykov and Davies, 1982).

Home-school collaboration. Although giving lip service to parental involvement, many school teachers and principals prefer to exclude parents from intimate involvement in the educational process. However, these barriers are crumbling in some schools, with special attention given to educational needs of low-income neighborhoods where school achievement has lagged. These schools are helping "parents act in educational capacities as home tutors, monitors of homework and attendance, guides for their children in the use of community educational resources, or engaging in other home activities to improve student learning." They seek greater interaction with the school, including "parent-teacher conferences, parents working individually with teachers, parent education classes, the use of indigenous outreach workers, phone contacts, and special activities for parents such as school programs, trips and cultural events" (Moles, 1981:10; Moles et al., 1982; also see Henderson, 1981).

In the Martin Luther King Elementary School in New Haven, serving a low-income black population, this thrust has extended into developing a support system for parents and their children within the neighborhood. Cosponsored by the New Haven Public Schools and the Yale Child Study Center, this program rests upon the assumption that parents, teachers, and principals all want children to succeed but that societal and institutional barriers have prevented positive collaboration. The project worked with school staff and parents separately to prepare them to work together more effectively. This has led to three levels of parental participation. The majority of parents have participated in broad-based social activities, fundraising events, report card conferences, and other conferences with school staff. From 10 to 40 percent of the parents have been involved in parent-staff workshops, committees, performed volunteer work at the school, or served as tutors. One to ten percent of the parents have worked with teachers and administrators in making policy decisions on curriculum and operating policies. Although research with a control group hasn't been conducted, in standardized tests the children at King School have ranked ahead of all other elementary schools in New Haven receiving federal compensatory education funds. Furthermore, the program has stimulated a number of parents to return to school for more education, find or upgrade their jobs, start business enterprises, and participate more in civic and political activities (Comer, 1980; 1982).

COMMUNITY EDUCATION

Another important neighborhood focus is the use of school buildings by the broader community. Community education, as this emphasis is called, encompasses six features: (1) an educational program for school age children, (2) maximum use of school facilities, (3) additional programs for children and youth, (4) programs for adults, (5) delivery of community services, and (6) community development through community councils (Minzey, 1981:278-280). By 1977, more than 1,000 school districts were operating over 5,000 community schools along these lines. School buildings are used throughout the day and evening and on weekends by a broad age range of the population. Some of them also provide space for community service workers, though rarely do they have enough room to be a full-fledge multiservice center. They tend to be organized so that regular school activities are kept separate from the after-school program, and they don't necessarily emphasis home-school collaboration. They may have advisory councils, but rarely do community schools provide for neighborhood control.

DEEMPHASIS OF NEIGHBORHOOD SCHOOLS

While many trends in public education stress the importance of community participation within neighborhoods, other trends and events have led to deemphasis of neighborhood schools.

Busing. To achieve racial integration, some school systems, on their own initiative or under court order, assign pupils to schools outside their neighborhood and if necessary provide bus transportation. The same thing happens where neighborhood schools are closed because of declining enrollment. Some districts use magnet schools with attractive, special features in order to promote voluntary desegregation. The latter are generally well accepted because of their voluntary nature, but in many places strong community opposition has arisen to closing schools, sending pupils outside their neighborhood to promote racial integration, and bringing children of another race into the neighborhood (for instance, see Buell, 1982 on the Boston experience). In these cases, we have a clash of social values: neighborhood versus economy and efficiency (for school closings) and the small community versus societal emphasis upon racial equality and integration.

Private schools. A variety of private schools function as other than neighborhood institutions. This is true of nonsecular private schools drawing pupils from more affluent families from a fairly wide area, and also of "alternative" schools formed to meet special needs, try out new approaches, and otherwise operate differently than the public schools. Parochial schools may serve mainly members of the sponsoring religious group, though many of them now enroll quite a number of nonmembers, including racial minorities who are dissatisfied with public schools. As this occurs, they become more like a neighborhood school, and some are setting up school advisory councils to achieve greater parental and community involvement. Here and there public elementary schools located in Catholic neighborhoods may be serving only a small percentage of school enrollment and may have relatively few ties to the neighborhood.

Analysis

The Yin-Yates survey of decentralized urban services (1974) encompassed human services and education as well as municipal services. They indicated that improved services occurred in a decided majority of cases reviewed. From a recipient neighborhood's perspective other gains from decentralizing human services are convenience, increased employment opportunities for residents, and greater sensitivity to localized needs on the part of the service deliverers.

The move toward the neighborhood hasn't been quite as wrenching for social services, health centers, and mental health programs because frequently the neighborhood programs have represented fresh starts with new federal funding rather than transfer of personnel and existing resources. However, struggle over control of schools has produced controversy in a number of cities as citizen activists challenged entrenched bureaucratic and union power. Efforts of multiservice centers to achieve coordination of human services has been as arduous as similar efforts to coordinate basic municipal services. Lots of persuasion, seasoned with some clout, and sweetened with extra funds makes up the recipe for success, but it's usually a struggle. It may be worth the effort, though, for it achieves horizontal integration of interrelated services, necessary to respond to complex needs. Likewise, each cluster of neighborhood services needs to be related to broader service systems and the specialized services available serving a wider population. This provides the vertical element of service integration.

Resident participation hasn't always come easy for the professionals: social workers, physicians, public health specialists, mental health practitioners, principals, school teachers. Fifteen years ago at the height of the protest era, citizens weren't always easy to deal with either. But time has healed wounds, actually working together has developed trust and mutual respect, successes have built confidence, and the general tone of society has become less abrasive (at least until the recession of the early eighties and federal cutbacks under President Reagan heightened anxieties).

Continued working together has made power less of an issue and has produced greater emphasis upon cooperation and partnerships. One of the most encouraging developments has been the recognition of natural helping networks. They've always been there, but in the last ten years they have come to be tied more closely to professional helping networks (Caplan and Killilea, 1976; Collins and Pancoast, 1976; Gottlieb, 1981; Froland and others, 1981; D. Warren, 1981). The concept presented in Figure 13.1, drawn to apply to community mental health, is equally applicable, by changing the names of some of the professional and lay participants, to youth development, delinquency prevention, restoration of offenders, home-school collaboration, services for the aging, and programs for a variety of other human needs. Rather than being a zero-sum game where one side loses when the other gains, professional and natural helpers working together boost the achievements of both. This is most likely to occur at the intimate neighborhood scale.

This idea of mutuality is also applicable in public education in situations where social policy has caused deemphasis of territory, specifically through busing for purposes of racial integration or efficient use of school facilities. With children from two or more neighborhoods brought together, their parents collectively form a community of interest, bound by their common concern for their children's education. Regretfully other concerns and fears have predominated in many instances — not wanting young children to have travel inconvenience, to go to a neighborhood perceived as unsafe or undesirable, or to mingle with another race; or not wanting those "others" to come to their neighborhood. In these cases the best methods of home-school collaboration need to be called upon so that parents from all segments are in communication with one another and with school personnel. In this manner, they can identify their shared concerns, deal with fears and hostilities, and work together for common goals. In short, many attributes of a good neighborhood can apply to the community of parents, teachers, and pupils drawn from several neighborhoods.

Exercises

(1) Choose one of the major functional areas of this chapter (social services, health, mental health, or education). Make a list of needs people have in that area and the kinds of services which respond.

(2) Determine which agencies and organizations in your city provide these services, their facilities, the kind of personnel they use, and where they are based. Which are located in neighborhoods and which, if any, are neighborhood-controlled?

(3) For neighborhood level services, determine the various roles played by residents.

(4) How are agency services related to natural helping networks, if at all?

(5) How is service integration achieved (a) horizontally within the neighborhood and (b) vertically within the service system?

(6) Do you think it is really possible to find a community interest among professionals, lay helpers, and citizens, or between parents from diverse racial and ethnic backgrounds, and then to act positively on common concerns? What are the obstacles? How can they be overcome?

CHAPTER 14

PHYSICAL PRESERVATION AND REVITALIZATION

As we reviewed in Chapter 7, neighborhoods change over time, both in occupancy and physical condition. Some are stable for many years with a constant population composition and persistent structural preservation. Others experience gradual or rapid population change, slow or accelerating deterioration, or reversal of decline and steady physical renovation. In this chapter we review methods used to preserve sound neighborhoods and to revitalize declining ones. We'll also touch on how to deal with severe decline.

Preservation and revitalization occur because of decisions made by numerous individuals, personally and in institutional roles. Roger Ahlbrandt and James Cunningham have offered a list of principal actors: (1979:26-27): local government, financial institutions, other institutions and organizations functioning in the neighborhood, appraisers, realtors, business persons, investors, homeowners, renters. Other important actors are landlords, owners of commercial properties and vacant land, developers, and potential homebuyers and renters. Local government actors hold a variety of positions, and some state and federal officials have significant roles. Indeed, there is more direct federal involvement in the physical aspects of neighborhoods than any other feature. (Also, see Downs, 1981:62.)

New Developments

Although this chapter concentrates on measures to preserve stable neighborhoods and revitalize wear-torn ones, we should also say a word about the development of new neighborhoods on vacant land, for this process affects the vitality of older neighborhoods. In this regard, Anthony Downs has pointed out that neighborhood changes occurring these days in large American cities are rooted in the urban development process that emerged following World War II to cope with explosive growth in metropolitan population. As he explained (1981:37),

> This process had to furnish moderate- and middle-income households with larger and higher-quality housing as their families grew and their real incomes rose. It had to expand tremendously the total number of

housing units available, while improving the quality of the housing inventory. Since it had to operate largely through markets where choices by individual households were voluntary, it had to cater to certain strong desires held by those households — to own free-standing single-family homes, to own automobiles, to live in neighborhoods that were relatively homogeneous socially and economically, and (among whites) to live in predominantly, and often exclusively, white neighborhoods. Not all these desires are commendable, but they were all strong enough to influence metropolitan-area growth patterns.

The bulk of new housing in the United States is privately produced and privately owned. During the last 40 years most new private housing has been priced for persons in the higher income range, seldom below middle income (unless a government subsidy was involved) and often higher. For most buyers their new homes are more expensive than the old ones, and they tend to sell to persons of a slightly lower income, who themselves may be upgrading their housing by moving from less expensive dwellings. Through a succession of sales, a large portion of the housing supply filters down over the years to lower-income households. However, this process occurs at different rates for different kinds of housing in various locations. When there is a large volume of new construction compared to demand, as occurred in the 1960s, the filtering process accelerates and many houses of lowest quality are likely to be vacated and abandoned.

In short, the housing supply and physical condition of all neighborhoods are affected by nature of the new housing market within the metropolitan area.

Preservation

The majority of American neighborhoods are in sound condition. The challenge for residents is to preserve their neighborhood, even in face of shifting populations and economic conditions.

INGREDIENTS

Successful neighborhood preservation has a number of ingredients; (HUD, 1975; McNulty and Kliment, 1976; Cohen, 1978; HUD 1979a).

Household income. It costs money to maintain housing, make mortgage payments, and pay taxes. Owner-occupants draw upon their personal income, and landlords use rent for these purposes. Although households have some flexibility in their budgets, below a certain income level they are hard pressed to meet other vital needs and pay what it costs, either as owners or renters, to pay for proper upkeep. Thus, what comes to public attention as a housing problem may be an income problem at its roots.

Desire and commitment. Even if they have sufficient income, property owners must have a desire and a commitment to keep their buildings in good condition. For owner-occupants this is a matter of pride and investment sense. Group psychology is a factor, too, that is, whether neighborhood peer pressure pushes for good maintenance and whether residents collectively have faith in their neighborhood's future. For absentee owners, investment wisdom can encourage effective maintenance, also personal pride, and (if necessary) enforcement of housing code regulations and pressure from tenants and citizen organizations.

Financing. Because most owners obtain mortgages to buy their properties, a sufficient supply of mortgage funds at terms people can afford is an essential ingredient for neighborhood preservation. Also, loans for housing improvements and neighborhood commercial ventures. We'll return to this need in a moment when we consider how neighborhoods can cope with disinvestment.

Social fabric. Vitality in the social fabric is another important ingredient. Ahlbrandt and Cunningham, based upon interviews with residents in Pittsburgh neighborhoods, have emphasized the following factors (1979:199):

(1) Personal relationships within the neighborhood, as present in kinship networks, friendship networks, neighboring, and common activity within clubs, organizations, and institutions.
(2) Interactions between and among clubs, organizations, and institutions.
(3) Interrelationships of families and individuals with suppliers of goods and services.
(4) Information flows (neighborhood newspapers, newsletters, etc.).
(5) Sharing of common values.

Subsequently they have added a sixth factor: conscious sharing of a common history.

Much of the social fabric is woven by informal relationships, but organizations consciously directed to neighborhood preservation and improvement have important contributions to make.

Coping with change. Because many neighborhoods experience shifts in population makeup over the course of years, they need the capacity to cope with change, including inmigration of people who are racially or ethnically different than the majority of the present population. Due to the arbitrage phenomenon we examined in Chapter 7 (p.100), sometimes new minority group residents have higher incomes than the persons they have replaced. Therefore, from an economic viewpoint there is no reason to expect lower maintenance. In fact, the opposite may occur. But when a lower-income population begins to move into a neighborhood, it may be necessary to pursue strict housing code enforcement to prevent lower maintenance and overcrowding and to add employment and other income programs to the roster of neighborhood services so that the new residents can afford to pay for housing upkeep.

Commerical enterprises. Shopping areas and other commercial enterprises are important features of many city and older suburban neighborhoods. They need preservation as much as the houses. Market forces provide the foundation, but organization of merchants to work together and provisions for adequate financing of their enterprises are also important, especially in view of the competition of regional shopping centers (for more on this topic, see Chapter 15).

Community services and facilities. Local government, private agencies, and institutions have important contributions to make to neighborhood preservation. Keeping up an adequate level of services and maintaining their facilities are essential. In neighborhoods with changing population they may need to adapt services and initiate new kinds of activities. The city, suburban municipality, and county can help strengthen the social fabric by providing meaningful resident participation in neighborhood planning, budget making, and other public activities affecting the neighborhood. Officially recognized neighborhood councils (see Chapter 16) can be strong tools for neighborhood preservation.

Regulations. The regulatory powers of local government can also be used in behalf of neighborhood preservation. Two important forms are the zoning ordinance, which indicates what land use, building types, and density of development may occur at specific locations, and the housing code, which sets standards for dwelling maintenance and occupancy. Traffic regulation is another important matter, for a neighborhood can be harmed by obtrusive commuter and truck traffic (Appleyard, 1981).

Partnerships. Many neighborhoods, particularly those well planned with relatively affluent and well-connected residents, seem to have the capacity to preserve their properties and way of life indefinitely. But others require concerted action which brings together residents, local government, and other interests in a working partnership (see Chapter 18).

OVERCOMING DISINVESTMENT

An enemy of neighborhood preservation is the withdrawal of capital by lending institutions and other investors. In Chapter 6, we cited a case of how one neighborhood in Chicago lost mortgage funds, housing maintenance jobs, local businesses, and the jobs they provide due to the disinvestment process. Other older neighborhoods have had the same experience.

Redlining. Among the severest forms of disinvestment is redlining, as practiced by lending institutions (HUD, 1978; Schafer and Ladd, 1981). Naparstek and Cincotta have explained (1976:8):

> Thus termed because more blatant practitioners draw red lines around target neighborhoods on area maps, redlining may consist of outright refusal to accept mortgage or home improvement loan applications. Or it may involve a number of subtler actions: awarding mortgage loans on inordinately short terms with high downpayment requirements; refusing to lend on properties older than a prescribed number of years; stalling on appraisals to discourage potential borrowers; underappraisals; refusing to lend in amounts below a fixed minimum figure, and charging inordinately high closing costs, to name a few.

As we saw in Chapter 9, citizen reaction against redlining in Chicago led to a vigorous campaign to get lending institutions to change their policies and eventually a national movement which produced two pieces of legislation. The Federal Home Mortgage Disclosure Act of 1975 requires financial institutions to disclose data on their lending records so that community groups and local governments can monitor performance. They can then pressure for greater reinvestment. The Community Reinvestment Act (CRA) of 1977 establishes that some 19,000 lending institutions in the United States have an affirmative responsibility to provide credit to their communities. The act ties this responsibility into bank examinations and federal approval of new charters, branch offices, mergers, relocations, and other banking activities. Community groups can enter their views at these points of decision making. A number of states and a few cities have also adopted disclosure laws and regulations.

Reinvestment strategies. Neighborhood organizations and coalitions in many cities have used these two acts as a means of pushing for greater investments in older neighborhoods. This experience has taught that a number of concrete steps are necessary to carry out a neighborhood-based reinvestment strategy (Fishbein and Zinsmeyer, 1980; HUD, 1980).

It starts by setting up a committee or task force to collect basic information about investment patterns, determine the roles of specific lenders, and assess the neighborhood's credit needs. Annual statements of lenders required by the Community Reinvestment Act are a valuable source of information. For many neighborhoods this analysis is likely to show that there are unmet, sound credit needs, that local lenders have capacity to meet these needs but aren't doing so, and that there are ways they can change their policies and practices.

Armed with this analysis, neighborhood organizations and coalitions can then enter into dialogue and negotiations with lenders about making greater investments in their neighborhoods. Among potential issues to discuss are the lender's overall lending policies and underwriting terms, the need for affirmative marketing, counseling for potential borrowers, provision of assistance to nonprofit organizations, and establishment of advisory committees and review panels to monitor progress. If negotiations fail, the organization can file a protest with federal regulators when the lending institution submits an application to branch, merge, change the location of a facility, or make other changes for which federal approval is required. The threat of such a protest can be a useful tool in the bargaining process.

Revitalization

There is a continuum among neighborhoods of those requiring preservation and those in worse condition where revitalization is necesary. Many of the same techniques apply to both, though more vigorous action and a greater input of new resources is necessary to achieve revitalization. Neighborhoods exhibiting a need for revitalization range from those with heavy deterioration to others where physical decline has just begun (Stegman, 1979; Housing Action Council, 1980).

Downs has noted that there are two types of revitalization: gentrification and incumbent upgrading and that a neighborhood may be affected by either or both. (For examples of combinations in St. Louis, see Schoenberg and Rosenbaum, 1980.) According to Downs (1981:72):

> In gentrification, relatively affluent newcomers buy and renovate homes in run-down neighborhoods. The poor initial residents are forced out. The newcomers are usually childless households — such as young professional couples. They make major investments of money and their own labor in upgrading their homes. Because of relatively high incomes, they are able to finance the improvements privately, usually with conventional mortgages.
>
> In incumbent upgrading, residents of run-down neighborhoods rehabilitate their housing themselves. Since they have relatively low incomes, they usually need assistance from publicly financed programs or subsidies.

PRIVATE RENEWAL

Some writers, though, use the term "revitalization" to apply only to the first type (London et al., 1980), which others describe as private neighborhood renewal (Clay, 1979) and some call the back-to-the-city movement (Michener, 1978; Laska and Spain, 1980). By whatever name it is known, the National Urban Coalition (1978:3), based upon a survey of experience in 44 cities, has identified four district phases of rehabilitation activity undertaken by these new owner-occupants: (1) a start-up phase initiated by a small group of pioneers; (2) a buy-in phase when confidence in the investment value of the neighborhood is growing; (3) a take-off phase when the area undergoing rehabilitation expands and new owner-occupants are joined by speculators looking for good investments; and (4) a fill-in phase when the remaining unimproved properties are purchased and fixed up.

Many city officials look favorably upon this process because increased property values mean higher property taxes, the new residents pay higher income taxes, and they require fewer social services and welfare assistance. Thus, some cities have encouraged private revitalization by increasing basic municipal services in those neighborhoods and improving community facilities.

DISPLACEMENT

In contrast, existing residents, their organizations, and outside advocates are bothered by gentrification because of the involuntary displacement it produces. The switch from renter to owner occupancy reduces the number of rental units, and in some cases may result in eviction. With a better market, landlords raise rents, and tenants who can't afford the higher rent have to leave. In addition, landlords pass on the higher taxes resulting from increased assessed value. Existing owners also experience higher taxes, and some who can't afford them reluctantly sell and move elsewhere. Elderly households and low income renters are hit hardest by the displacement process. (For more on displacement, see Weiler, 1978; Solomon, 1979; HUD, 1979c, 1981; Hartman, Keating, and Le Gates, 1982; Lang, 1982).

Another form of displacement occurs when rental apartments are converted into condominiums in which occupants own their own units. Not all renters want to own or can afford the down payments and monthly carrying charges, so they have to move. Condo conversion strikes hardest at the lower income elderly, who are forced to relocate but may have difficulty finding suitable alternative housing they can afford. This process occurs in gentrifying neighborhoods and in other parts of the city and suburbs attractive to higher income buyers.

PLANNING

Revitalization can be enhanced through careful planning involving residents, local government, and other interests.

Neighborhood planning. The city planning movement which started in the United States in the early part of this century tended to be grandiose in scope, but from 1929 onwards, influenced by Clarence Perry's advocacy of neighborhood planning units (see p. 54), city planners began to give some attention to residential neighborhoods. By the 1960s neighborhood planning was fairly common, using three different modes.

The first mode was handled by professional planners working at planning department headquarters where they undertook studies of land use, housing conditions, community facility needs, and transportation. When completed, the planners took their plans to neighborhoods for comment. In the second mode advocate planners with an independent base took the initiative, though seeking input from residents (see p. 130). The third mode built upon collaboration between a neighborhood advisory committee (sometimes elected) and neighborhood planners working for the city. The first and third modes produced neighborhood plans which were incorporated into the city's comprehensive plan, after hearings, review, and possible modification by the planning commission and city council. The product of the advocate planners formed the basis for negotiation with city officials. Presently most cities engaged in neighborhood planning use the third mode, though practices vary on how fully residents are involved. Neighborhoods with their own development organizations are in a better position to offer input.

A typical neighborhood plan, as produced by city planners, is likely to contain (1) an inventory of existing conditions including population characteristics, land use, housing, commerce and industry, community facilities, and transportation, (2) discussion of neighborhood problems, (3) specification of goals, objectives, and strategies, (4) proposed land uses and zoning, (5) recommended neighborhood projects and schedule for implementation, and (6) estimates of costs and resources required (Urban Systems Research and Engineering, 1980; also see Werth and Bryant, 1979 and Hallman, 1981c).

Project planning. A condensed version of the neighborhood planning process has occurred under the successive federal redevelopment, urban renewal, and community development programs, focusing more sharply on specific activities to be funded for specific projects. In the 1950s city planners and redevelopment staffs did most of the project planning with citizens reacting to completed plans. In 1968 the U.S. Department of Housing and Urban Development (HUD) mandated the use of project area committees in all urban renewal projects involving rehabilitation (most of them did), and some of these committees gained their own staff. In this same period advocacy planning provided an independent voice in renewal planning, and the Model Cities Program funded staff assistance for many of the participating neighborhoods. Under the Community Development Block Grant (CDBG) Program, enacted in 1974, most cities with neighborhood projects make use of neighborhood advisory committees or neighborhood councils. Some neighborhood development organizations and community development corporations have prepared their own plans.

PROGRAMS

Neighborhood revitalization plans, especially those aimed at incumbent upgrading, are implemented through a variety of programs, financed with public funds and private investments, often in combination (Kollias, 1977; Cassidy, 1980). At an earlier period, neighborhoods in the worst conditions, and some not so bad, were cleared wholesale and redeveloped, usually with more expensive housing or a different use. Nowadays clearance occurs more sparingly and much more attention is given to improving existing structures.

Housing rehabilitation. Accordingly, housing rehabilitation plays a major role in most revitalization programs. Experience of 35 years has shown that to succeed, programmed housing rehabilitation must combine competent workmanship, sufficient financing, enforcement of basic housing code standards, and support from tenants and other residents for proper upkeep. It is advisable to fix up all the dwellings in a block or a subsection of the neighborhood at the same time to achieve reinforcement and safeguard investments. Homeowners may need technical assistance in deciding what to do, choosing contractors, and arranging financing. Neighborhood development organizations and private companies both have roles to play. Lenders and federal assistance programs must be mobilized. Some neighborhoods have used workers assigned by public service employment programs, and some rehabilitation projects have incorporated sweat equity, enabling future owner-occupants to use uncompensated labor as a down payment. Some programs have included energy conservation measures. These varied techniques have been applied successfully to a wide range of buildings, including some which would have been candidates for demolition 30 years ago (NAHRO, 1979).

Construction of assisted housing. Because many city residents can't afford to pay private market rent for decent housing, a variety of publicly financed programs have evolved. Beginning with the Housing Act of 1937 the federal government has provided subsidies to local housing authorities for the construction and operation of low-rent public housing. New projects are no longer being constructed, but the older ones are present in older neighborhoods and many are themselves candidates for rehabilitation. Private developers and nonprofit corporations have built new housing aided by federal mortgage insurance, rental assistance, and direct grants. A considerable number of states and a few cities have adopted their own assisted housing measures, including property tax abatement. Federal tax laws offer tax shelters to wealthy persons who invest in certain types of housing. Some insurance companies and other large institutional investors have made mortgage loans for rehabilitated and new housing, undertaken as part of neighborhood revitalization projects. Some of the more sophisticated neighborhood development organizations have been able to leverage funds from a variety of sources to carry out their housing activities.

Homesteading. In the early 1970s mortgage foreclosures in federal programs and local tax foreclosures yielded a considerable number of vacant houses basically in sound condition though requiring some rehabilitation. Wilmington, Delaware, followed by Philadelphia and

Baltimore, instituted homesteading to sell the vacant buildings for a nominal price (such as $1.00) to persons who would agree to fix them up and live in them for at least a minimum period (such as four or five years). The idea caught on, and in 1974 Congress appropriated funds to extend the program to other cities. Elements of a homesteading program include selection of the properties to be sold, choice of homesteaders, a rehabilitation plan for each structure, technical assistance to the homesteader, and arrangements to finance repair costs (Hughes and Bleakly, 1975; Urban Systems Research and Engineering, 1977).

Because some cities and HUD in some places have been slow to make available vacant properties for homesteading, neighborhood activists have occupied buildings as squatters and initiated rehabilitation, usually with sweat equity and with financing pieced together however they can arrange it.

Historic preservation. Many older neighborhoods have considerable historical value because of both architectural and cultural heritage. Thus, historic preservation is an added reason for neighborhood revitalization. To the basic techniques of housing rehabilitation is added a special emphasis on design integrity, possibly reinforced by a local ordinance prohibiting demolition or improper alteration of historic structures. Some historic neighborhoods have attracted upper income households, leading to gentrification, but some localities have taken steps to preserve an income mixture through regulatory and financing devices.

Community facilities and services. Older neighborhoods are likely to have outmoded streets, sewers, and public buildings. By replacing or improving them, the city can reinforce private investments and the inflow of federal and state housing funds. Services designed to meet the needs of residents, such as those considered in the two previous chapters, are also important.

Economic development. We are saving the topic of neighborhood economic development until the next chapter. Many of the activities considered there — commercial revitalization, capital infusion for new and existing enterprises, job creation activities — are proper components of neighborhood revitalization.

Social fabric, resident participation. Our remarks in the previous section on neighborhood preservation regarding the importance of strengthening the social fabric and involving residents in planning and implementation apply equally to neighborhood revitalization.

REGULATIONS

In addition to specific programs, regulatory measures can contribute to neighborhood revitalization.

Zoning, housing code. The zoning ordinance and the housing code (previously mentioned) are adopted and administered by city government. In this way the city council determines what land uses are allowed in each neighborhood, and zoning boards of adjustment determine what special exceptions are permissible. In the strictness or laxness of housing code enforcement, the city administration affects the quality of neighborhood housing conditions. In this manner, zoning decisions and housing code strategies have significant effects upon neighborhood stability, decline, or revitalization.

Under the Model Cities Program (1967-1974) there was some experimentation with neighborhood land use boards, such as in Seattle, but no city neighborhood has succeeded in gaining this power fully. (Neighborhood-size suburban municipalities, however, have their own zoning ordinances.) Nevertheless, cities with official neighborhood councils automatically refer all proposed zoning changes affecting their neighborhoods to them for comment, and some cities do likewise with neighborhood associations registered to receive notices. Even where not notified, neighborhood organizations can discover prospective zoning changes and use interest group tactics to influence decisions on them.

Dealing with displacement. Some cities have adopted regulations to deal with condo conversion and other forms of displacement (Solomon, 1979; Hartman et al., 1982). A few places have rent control to limit the rate of increase, usually pegged to cost of operations and a reasonable return on the owner's investment. Many more locales have eviction controls, requiring adequate notice and just cause. Condo conversion controls might require that a certain percentage of tenants must consent to conversion, that tenants have first right to purchase their own units, that the condo converter must pay the relocation costs of persons displaced, that a portion of the units be reserved for continued rental occupancy by current elderly tenants at the existing rent level (Silver and Shreve, 1979). In a few cities the tenants have the first right to purchase the building and operate as a cooperative. Some tenant groups have done so by obtaining federal housing assistance under low- and moderate-income programs. Although these various regulations are citywide in scope, neighborhood organizations use them to cope with displacement occurring in their neighborhood.

ORGANIZATIONS

As noted at the beginning of the chapter, many actors have roles in neighborhood revitalization. Here we highlight two varieties of neighborhood-based organizations which have emerged in recent years.

Neighborhood development organizations. Since the 1960s a considerable number of neighborhood organizations have incorporated and entered into housing and other developmental activities. They work mainly in low- and moderate-income neighborhoods and usually receive some kind of federal assistance (Schur and Sherry, 1977; Bowsher, 1980).

A good sample of who they are and what they do comes from a survey conducted by the Urban Institute of 99 neighborhood development organizations (NDOs) which received funds under the Neighborhood Self-Help Development Act of 1978 (Marshall and Mayer, 1983). These 99 groups actually represent several categories: nonprofit housing organizations (46), tenants' groups (3), neighborhood organizations with other single purposes (11), multipurpose organizations (31), and coalitions (8). Seventy-eight of them had incorporated within the last ten years, and 42 were less than six years old. Thirty-eight had started as advocacy organizations and switched to development, while 44 focused on development from the beginning; 12 were original human service organizations, four started as community action agencies in the 1960s, and one was multipurpose from the beginning. The median annual budget was $250,000 and ranged from 15 organizations with less than $50,000 to 19 organizations with over $1 million.

This study determined that the NDOs had made very substantial progress in carrying out their development projects, even in face of rising interest rates, declining public funding, and a worsening economic climate. They succeeded in obtaining project funds from a variety of sources, were effective in promoting self-help, and systematically directed project benefits to neighborhood residents and businesses. The researchers judged that the following internal characteristics, residing in the NDO staff and board, were the most significant contributors to good performance (Marshall and Mayer, 1983:13):

- A broadly skilled executive director.
- A key staff member with experience in and knowledge about development projects.
- An understanding of issues of project financial feasibility and their importance.

- A track record of accomplishments in development and related neighborhood work.

The most important aspects of NDOs' relationships with others were the following (1983:14):

- The level of support from its own community for an NDO's projects and overall efforts.
- The quality of NDOs' working relationship with local government, especially their ability to form a genuine cooperative partnership.
- Access to competent technical assistance in fields where NDOs lacked their own expertise.

Neighborhood housing services. The partnership factor is a key ingredient in another variety of organization, neighborhood housing services (NHS). The first NHS began operations in Pittsburgh in 1968 with local funding (Ahlbrandt and Brophy, 1975). The idea came to the attention of William Whiteside and some other executives in federal agencies which regulate lending institutions and became the primary interest of the interagency Urban Reinvestment Task Force, which formed in 1974 with Whiteside as director. The Task Force worked with local government officials, lenders, and neighborhood representatives to establish neighborhood housing services in other cities. In 1978 Congress converted the Task Force into the Neighborhood Reinvestment Corporation. By the spring of 1983 there were more than 185 NHS operations in 134 cities throughout the land.

A neighborhood housing service is organized as a private nonprofit corporation and is governed by a board of directors consisting of representatives from the neighborhood, lending institutions, and local government, with the residents having majority membership. As a minimum the staff has an executive director, housing rehabilitation specialist, and administrative assistant/secretary, and maybe others. NHS staff work with property owners to achieve housing improvements. The participating lender assure that adequate financing is available, sometimes through a mortgage pool. The NHS also has a revolving loan fund for owners who can't obtain financing through conventional channels. The city takes responsibility for housing inspection and code enforcement and undertaking public improvements in the neighborhood. NHS funding comes from a combination of grants from the Neighborhood Reinvestment Corporation, money raised locally by lenders, CDBG funds allocated by the city, and sometimes the city's own funds (Urban Systems Research and Engineering, 1980).

In an evaluation of a sample of a dozen neighborhood housing services, Phillip L. Clay concluded that their partnership approach was succeeding in achieving reasonably quick rehabilitation, supported by cooperation from lenders (who may have been neglecting the neighborhood previously), city agencies, and the residents themselves. "While the rehabilitation is in only a few cases extensive," he noted (1981:169), "the physical results are in most cases significant and visible." Others have pointed out that the NHS model is applicable only in certain neighborhoods where home ownership is fairly high (though the Reinvestment Corporation has experimented with a multifamily dwelling model), but that where it is used, it can successfully achieve its objectives.

Dealing with Severe Decline

Among the greatest challenge to neighborhood revitalization are neighborhoods in an advanced stage of decay, particularly those where considerable abandonment has occurred.

Abandonment. During the late 1960s housing abandonment reached epidemic portions in some city neighborhoods in the northeast and the industrial midwest (Sternlieb and Burchell, 1973). Partially this resulted from a sustained housing boom and some lessening of racial discrimination so that moderate and middle income persons living in older, deteriorating neighborhoods could find better quality housing elsewhere. Although some poor people still lived in overcrowded quarters, there was a net reduction in demand for low quality housing. Landlords found maintenance and operating costs rising (including higher fuel costs after 1973), but not enough potential renters who could afford to pay higher rents. Vandalism, rent withholding in protest of poor maintenance, and general neighborhood deterioration added to the problem. Owners experienced a negative cash flow but couldn't find buyers for their properties. They might also be in arrears in property taxes. So they vacated the building and abandoned it. At the same time, city government might reduce services, or at least not keep up with the greater needs of the poor neighborhood. As one study pointed out (HUD, 1973:8):

> In some neighborhoods the point is reached where the neighborhood-wide extent of physical deterioration and poverty and municipal neglect become so pervasive that a new phenomenon comes into existence: concentrated and contagious abandonment. When this occurs, the

problem shifts from an individual building problem to a neighborhood problem where the conditions of proverty and the almost complete withdrawal of capital create a very undesirable living environment.

Rebuilding. In the past, two approaches to this condition have been applied. One way was to let abandonment run its course until private entrepreneurs bought the vacant land and then constructed new buildings, often for commercial and industrial uses rather than new housing. The other way was public acquisition through redevelopment followed by new construction, which might be upper income housing if the location was suitable, subsidized housing for lower income groups, commercial development, or industrial use.

During the 1970s the U.S. Department of Housing and Urban Development sponsored research and public forums to consider adoption of a combination of these approaches under the triage principle. The term "triage" is French for "sorting." Battlefront medics use triage to divide the wounded into three groups: those who will survive without much medical attention, those for whom prompt attention will likely save their lives, and those who are so severely wounded that survival is highly unlikely. The medics concentrate on the second group and give the third sedatives to reduce their suffering as they die. Likewise, so the argument went, neighborhoods can be divided into three groups: healthy areas, very deteriorated areas, and those in-between. The latter would receive the preponderance of available community development resources, and "there would be no large expenditures for major upgrading effects in most parts of very deteriorated areas" (Downs, 1975:18-24; also see Public Affairs Counseling, 1975). This proposal evoked an emotional debate about government turning away from severe need, and HUD pulled back from this doctrine. Nevertheless, without saying so, some cities practice triage through inaction.

Elsewhere, neighborhood organizations in some severely deteriorated areas have refused to let their neighborhood die. They have taken over abandoned buildings, initiated self-help rehabilitation and drawn in public funds, such as in New York's Lower East Side (Schur, 1979). In some places city government has joined in and is making available funds for new construction and supportive activities. The best of these efforts are dealing not only with physical reconstruction but also with the need for human resource development and are using construction work for job training and employment of residents.

One of the strongest efforts built around public/neighborhood/ private sector cooperation is occurring in the South Bronx in New York City, an area where wholesale abandonment occurred during the 1970s.

Programs encompass housing rehabilitation and new construction, economic development featuring industrial and commercial projects for job creation, employment training, human services of many varieties, and land revitalization to turn rubble-strewn land into parks and building sites. Involved are city, state, and federal agencies, numerous community organizations, churches, social welfare agencies, and the private business sector. To pull these elements together, the city created a nonprofit organization, the South Bronx Development Organization, with a board consisting of city and state officials and representatives of six community boards. Although the South Bronx still faces enormous problems, the simultaneous application of physical, economic, and social programs has reversed the pattern of decay and abandonment.

CONCLUSION

The South Bronx experience, verified in numerous neighborhoods of many kinds in other cities, indicates that the passage through stages of decline and revitalization can move in either direction. Physical death isn't the inevitable fate of a neighborhood, no matter how old and seemingly obsolete its structures are. The redevelopment strategy of the 1950s, featuring relocation of persons from occupied buildings and wholesale clearance of the structures, is definitely outmoded. Of course, some demolition may be necessary for structures deteriorated beyond the stage of economic repair. But rebuilding can be achieved in ways to minimize displacement and to open new opportunities for the present residents. Involvement of residents through their own organizations is essential, but also both government and the private business sector must play important roles. Physical rehabilitation and reconstruction alone is insufficient, for the social and economic problems must also be addressed.

This same need for multiple strategies is equally true for neighborhoods in good condition and in an early stage of decline. Program elements will differ but a concern for all elements of neighborhood life should be part of every effort of preservation and revitalization.

Exercises

(1) In your city identify two neighborhoods, one in good condition where preservation is needed and another in worse condition where revitalization is required.

(2) Write a description of physical conditions in each neighborhood and the major problems.

(3) What programs are now going on to solve these problems? Are they succeeding?

(4) What is the relationship between physical, social, and economic programs?

(5) What are the roles of neighborhood organizations, government, private enterprise, and voluntary agencies from outside the neighborhood?

(6) What program elements and actors seem to be missing which would enhance chances for success?

CHAPTER 15

ECONOMIC DEVELOPMENT

As we discussed in Chapter 6, a neighborhood can be perceived as a little economy. It is the site of economic resources: wealth in the form of land, buildings, and equipment and the human resources of labor, brainpower, technical and entrepreneurial skill, and will. There is a flow of money, goods, services, and people engaged in economic pursuits into, out of, and within the neighborhood.

Because a substantial part of our economic life occurs in a city and metropolitan context — for work, shopping, banking, insurance, buying automobiles, and transportation, most of us don't pay much attention to the minieconomy of our neighborhood. However, some persons and organizations do, especially in low and moderate income areas, motivated by a desire to strengthen the neighborhood economy. Their efforts are the focus of this chapter as we look at neighborhood economic development activities and organizations which have emerged during the past 20 years. We'll give most attention to community development corporations, which have multifaceted activities, and to commercial revitalization programs, but we'll also touch on a variety of other approaches. The housing activities we examined in Chapter 14 have economic features and should be kept in mind, too. We'll consider mostly the summit of achievements; realizing that many neighborhoods have done very little in economic development and relatively few have adopted a comprehensive approach.

Community Development Corporations

In the 1960s a new instrument emerged to promote the economic development of low-income areas: the community development corporation (CDC). This term identifies "organizations created and controlled by people living in impoverished areas for the purpose of planning, stimulating, financing, and, when necessary, owning and operating businesses that will provide employment income and a better life for the residents of these areas" (Faux, 1971:29). CDCs take many forms and have different emphases, but, according to Geoffrey Faux, they have the following common characteristics (1971:30):

Economic development: CDCs are engaged in all stages of economic development. They act as planners, provide technical assistance to local entrepreneurs, make investments and operate businesses.

Local control: CDCs are controlled by residents or by representatives selected by residents of the poverty-stricken areas they seek to serve.

Social goals: Their ultimate purpose is to increase the economic well-being of neighborhood residents.

Most CDCs are nonprofit corporations, but many of them have for-profit subsidiaries. In some cases, the parent corporation is profit-making and has one or more nonprofit subsidiaries. Most of them are structured in the standard hierarchical mode of business corporations with a president or executive director reporting to the board of directors and taking charge of staff operations. Depending upon program emphasis, staff might include economic development planners, business and finance specialists, community development personnel, and managers of housing, employment, and other program divisions. Community control comes through membership or stock ownership. Members or stockholders elect some or all of the board of directors, though some board members may be selected by designated organizations. The board includes residents and business persons from the neighborhood and often a few business leaders from the broader community (National Commission on Neighborhoods, 1979:140-148; Kelly, 1977).

ORIGINS

Community development corporations emerged out of the seething discontent of black and Hispanic communities, driven by a desire of the initiators to create jobs and business opportunities. Thus, in North Philadelphia in 1962 Rev. Leon Sullivan got members of his church to contribute to and invest in Progress Enterprises, which combined a nonprofit charitable trust and a profit-making holding corporation. Later other community residents became shareholders. This provided the capital for investments in housing, an in-town shopping center, an electronics plant, a garment manufacturing enterprise, and other commercial ventures (Sullivan, 1969; Garn et al., 1976, 71-96).

In the sprawling south central area of Los Angeles, labor leader Ted Watkins organized the Watts Labor Community Action Committee (WLCAC) in the spring of 1965. Initially, the United Auto Workers (UAW) paid his salary. After the Watts riots that summer, federal agencies and foundations began pouring money into this neglected area, much of it channeled through WLCAC. Job training and employment programs were the initial thrust, but from 1968 on economic development received increased attention (Hallman, 1974:159-162). In the same metropolis the UAW, the Ford Foundation, and several federal agencies joined together in 1968 to assist the formation of the East Los Angeles Community Union (TELACU) in a predominantly Mexican-American community (Faux, 1971:79-80).

Two organizations which started in the early 1960s with Saul Alinsky's assistance, the Woodlawn Organization (TWO) in Chicago and FIGHT (Freedom, Independence, God, Honor, Today) in Rochester moved from confrontation and protest into economic and community development programs (Brazier, 1969; Fish, 1973; Garn and others, 1976:43-70). In Cleveland, Deforest Brown, a black minister, established the Hough Area Development Corporation. In 1966 Senator Robert F. Kennedy stimulated the formation of the Bedford-Stuyvesant Restoration Corporation and convinced IBM and other business interests to get involved (Stein, 1975; Garn and others, 1976:11-41).

Kennedy joined with Senator Jacob Javits in sponsoring a new federal program to assist such community initiatives. Known as the Special Impact Program, it was originally administered by the U.S. Department of Labor but shifted to the Office of Economic Opportunity in 1986. There it grew until OEO was providing administrative, planning, and investment funds to approximately 40 urban and rural CDCs. The Ford Foundation also gave substantial support to eight CDCs, with some overlap (Ford Foundation, 1973). In that same period the Model Cities Program, run by the U.S. Department of Housing and Urban Development, also helped fund CDCs. Other forms of federal assistance became available.

In 1974 OEO was reorganized into the Community Services Administration and the Special Impact legislation was refined into Title VII of the Community Services Act. In that same year the Community Development Block Grant (CDBG) Program replaced Model Cities, spread the money to more neighborhoods, but permitted cities to support some CDC activities (especially after a 1977 amendment). Meanwhile, other public and private funding sources have come available, though waxing and waning through the years. With outside funding has come technical assistance provided by various federal, state, and local governmental agencies, national and regional nonprofit organizations, and private consultants.

By one estimate there were 700 community-based economic development organizations around the country at the end of the 1970s (Litvak and Daniels, 1979). Since then, cutbacks in federal funding have inhibited their work, leading to termination of some weaker CDCs and diminished activities for many of the others.

ACTIVITIES AND ACCOMPLISHMENTS

What community development corporations do and what they have achieved is illustrated by an evaluation of the accomplishments of 15 urban and rural CDCs receiving Title VII support. Undertaken by the National Center for Economic Alternatives (NCEA, 1981), this study described five strategies pursued by CDCs: capital infusion, employment, service delivery, community institution building, and pressure group. CDCs usually combine several strategies, dealing with different aspects of the neighborhood economy. (Also see Mahmoud and Gosh, 1979; National Economic Development and Law Center, 1983)

Capital infusion. A basic premise of the CDCs is that their community has insufficient resources for business enterprise development, industrial plants, improvement in housing stock, educational and health facilities, and other community institutions. Their task, therefore, is to bring in new capital. Federal funds from the Special Impact and Title VII programs were the starting point, but they were able to draw in other public and private funds. Among the urban CDCs covered by this study, 89 percent of the business development activities were business ventures and 11 percent dealt with financial institutions. Of the business ventures, construction and property development drew the greatest attention (48 percent), including residential, commercial, industrial, land banking, and formation of construction companies. Manufacturing, retail or wholesale, and service business constituted 41 percent of the activities (NCEA, 1981:84).

Among the CDCs studied manufacturing ventures with markets outside the special impact area were the most likely to survive. Not as effective were those pursuing only local markets, having a protected market, or buying out local manufacturing firms which were in trouble. Realistic marketing studies were a key ingredient (NCEA, 1981:93-95).

In wholesale and retail trade, large ventures with a broader market did better than small ones with a narrow market. Association with a large corporation, such as a supermarket or fast food chain, helped survival. Service ventures with low capitalization and low-skill employment didn't do well, nor did joint ventures with individual and family enterprises. More success occurred where CDCs set up firms to provide management services to themselves and in other ways to have partially captive markets. In recent years some CDCs have assisted firms involved in radio, community cable television, and other communications enterprises, but these are too new to evaluate their accomplishments (NCEA, 1981:98-100).

Physical development and construction has received considerable attention from urban CDCs because the halt of deterioration and improvement of the physical infrastructure is necessary for revitalization of the impact areas. They have worked on maintenance and improvement of commercial space, suitable industrial facilities, roads, basic public utilities, and safe and sound residential areas. Most of their real estate activities are related to other community economic development projects (NCEA, 1981:105-111).

The CDCs have found it useful to help set up financial intermediaries, often under a special federal program. These have included small business investment corporations (SBICs) related to the Small Business Administration, minority enterprise small business investment corporations (MESBICs), local development corporations (LDCs), revolving loan and loan guarantee funds, wholly owned thrift institutions, credit unions, and equity in minority-owned banks. "These intermediaries," NCEA reported (1981:117), "performed more favorably which made investments that fit into an already well-articulated strategy for making direct business loans and equity investments in particular sectors." Nevertheless, "because of their size and often necessary reliance on private participation, capitalization of either partially or wholly CDC-controlled intermediaries of this kind has been difficult."

Indeed, all along the line the community development corporations have faced grave difficulties, for they are working in neighborhoods where the private sector has withdrawn investments. "Most of the CDCs have experienced serious project failures at one point or another," NCEA reported (1981:8). Some of the ventures assisted by the sample CDCs were no longer in existence at the time of the study and some of the remainder weren't profitable, but the overall bottom-line for all assisted activities showed a net profit. This led NCEA to conclude that "the Title VII program has been able to achieve the kinds of effects called for in the legislation and regulations: raising capital for the impact areas, providing a net inflow of assets, creating profitable business enterprises" (1981:80).

Employment. Usually part of the CDC strategy has dealt with the lack of employment opportunities and the low income and wage levels of the neighborhoods. They have responded by creating jobs through business and physical development and by helping to retain jobs already in the neighborhood. They have improved employability of the residents through training and social services and have helped them gain better access to jobs in the local and regional economy. They have also sponsored public service employment activities for the dual purposes of job creation and community improvement.

Service delivery. Where CDCs have become involved in service delivery, it has usually been as a support for economic development and employment programs. Projects have included job training, health service, and daycare.

Institutional development. The establishment of the CDC itself was the first step of institution building. They have had to develop planning capability, managerial skill, and political finesse. They have learned that economic development is a process requiring many years to accomplish, not merely a set of special projects. This requires both short- and long-term planning and comprehensive programming.

CDCs have created financial intermediaries and for-profit subsidiaries. CDC staff have served as a source of technical assistance to other organizations in the community. The core staff has been an orchestrator of resources. Indeed, the NCEA study observed (1981:90), "The special characteristics of CDCs appears to be their ability to broker, channel, link, and package resources to create new investments and institutions where there had been none previously."

Pressure group. One of their strengths has been their access to politically controlled resources. Some of them started as protest movements. All of them got going during a period when federal agencies, foundations, and business leadership wanted to do something about poor, minority communities. This has opened doors which were closed 25 years ago.

In the course of their evolution, most CDCs have settled into fairly calm operations. If their initiators were protest leaders, they have either modified their approach or been replaced by managerial types. Nevertheless, they remain community institutions, and they always have in reserve the potential for mounting protest or other pressure group tactics to assure that resources continued to flow to their communities. This is one of several elements in the mixed strategies pursued by community development corporations.

Commercial Revitalization

Another focus for economic development is older neighborhood commercial areas, usually developed as strips along major streets or as a cluster at the intersection of two major arteries. Once a vital force in both neighborhood life and the city's economy, many of these neighborhood commercial areas have suffered adversely from the competition of suburban shopping centers and sometimes from lowered personal income of the residents due to population changes. They have empty stores, vacant lots, inadequate parking, and perhaps increased street crime. Yet, in face of these difficulties, neighborhood commercial revitalization has occurred in many cities. From this experience, a number of national organizations working with local merchants and neighborhood leaders have extracted a body of practical knowledge on successful methods.[1] A good deal of this activity has focused upon white ethnic neighborhoods, but black and Hispanic areas have also been involved.

ORGANIZATION

"The structure for doing commercial revitalization and neighborhood development requires a triangular form of partnership between the private sector, public sector and the neighborhood groups"(NCUEA, 1979b:23). This can take a variety of forms. It may be desirable for merchants to form their own association to deal with common concerns and then create an umbrella organization involving themselves, representatives of neighborhood organizations and institutions, city government, and citywide business associations. Some neighborhoods have formed "local development corporations" in order to qualify loans from the Small Business Administration. Whatever the form, a partnership approach is essential because all these interests have a stake in commercial revitalization and a contribution to make.

Sometimes partnership is achieved through the creation of task forces, dealing with such matters as merchant assistance, membership and discipline, special events and promotion, cooperative advertising, new business development, financial mechanisms, physical and environmental issues, parking and traffic, protection and public safety, grievances and public relations, public services (NUDSC, 1977:19).

PLANNING

This task force list outlines the main items of the agenda for developing and implementing a plan for neighborhood commercial revitalization. Planning starts by analyzing what now exists and estimating the potential. This can include a survey of the physical characteristics of the commercial area (building size, usage, vacancies, and condition; parking; condition of streets, and sidewalks; environmental deficiencies). There can also be surveys of merchants' attitudes and plans and shoppers' and residents' preferences. This can lead to an analysis of the market area, consumer buying power, the competition, characteristics of present and prospective shoppers, and total potential market. Next comes the setting of goals and objectives and specification of implementing actions, including business development, physical improvement, financing, promotion, and marketing (NCUED, 1979b:4-11).

FINANCING AND TECHNICAL ASSISTANCE

Where commercial decline is occurring, merchants may have difficulty to obtain the financing they need for business operations and improvements. Individually they meet resistance from banks to make loans, but when they are part of a concerted improvement effort, they will have a better chance. Moreover, they may qualify for loans, guarantees, and other financial mechanisms sponsored by city, state, and federal agencies. It is desirable to have a financial packager on the local staff or as a consultant to help merchants tap available resources.

They can benefit from other kinds of individualized technical assistance, such as for merchandising, promotions, store layout, market analysis, and basic management. The merchants collectively also can use professional advice in how to organize and work together, integrative design features, joint advertising, special events, traffic and parking, and other common concerns.

Another part of a business development plan is to attract new businesses to occupy vacant stores and to offer goods and services which complement or reinforce existing businesses. Possibilities include ethnic restaurants, professional offices for persons being priced out of downtown by high rents, wholesale firms and discount stores serving multineighborhood markets, and new, incubating businesses looking for fairly low rent. Market analysis prepares the way, but the merchants association, the local development corporation, or some other organization must take the initiative and be available to offer technical assistance and guidance on sources of financing.

PHYSICAL IMPROVEMENTS

Most older neighborhood commercial areas need physical improvement of private properties and public facilities.

Some commercial strips have realized that their buildings are a rich architectural heritage, though often cluttered and distorted by individual changes which have occurred over many years. They may not be able to restore the original design completely, but building owners and leasees can work together to bring out the best in the architectural character. Fresh paint, cleaned and repointed masonry, harmonious design of signs and display windows, landscaping, and new street furniture can add to the attractiveness (National Trust, 1979).

Lack of parking and congested streets especially as compared to suburban malls, keep customers away. Therefore, city government has an important contribution to make in working with the merchants and residents to provide more off-street parking, repair streets and sidewalks, route traffic to help the commercial area without harming nearby residential streets, develop and maintain miniparks, and make other physical improvements.

Some cities have bond issues and operating funds assigned to improving neighborhood commercial districts. A few localities provide for tax-increment financing whereby the commercial area is treated as a special assessment district, the city borrows money to pay for current improvements, and pays off the bonds from increased tax revenues coming in because of commercial revitalization.

JOINT ACTIVITIES

Working together carries into the ongoing operations of the neighborhood commercial area. This is particularly needed for marketing and promotion, especially since the competing suburban malls are well equipped for joint endeavors. Among promotional tools are shopping guides, flyers, coupons, sidewalk sales, local celebrations (working with the neighborhood association), contests, banners, and window art (NCUEA, 1979b:28-30; National Trust, 1979:59-60). Staff is necessary to carry out such joint activities because individual merchants are too busy with their own businesses. Such staff can be an employee of the merchants association or a consultant.

Revitalized neighborhood commercial areas have different character than the downtown shopping district and suburban shopping centers, but they fulfill a vital function both for their neighborhoods and for the city as a whole.

Other Initiatives

During the past 20 years many other techniques have been used to develop and strengthen economic activities in lower-income neighborhoods. We highlight a few more.

MINORITY BUSINESS ENTERPRISES

Most of the urban community development corporations have functioned in black and Hispanic neighborhoods, and some of the commercial revitalization projects have, too. In the course of their work, they have strengthened existing minority-owned businesses and helped new ones to form. In addition, some national, state, and foundation programs have aimed particularly at minority business enterprise apart from neighborhood economic development. For instance, the Nixon administration took initiatives to encourage "black capitalism." These efforts use techniques similar to some methods of CDCs and commercial revitalization projects: Loans, guarantees, technical assistance, sometimes tax incentives. Because of patterns of business and residential location, many of the businesses assisted are in minority neighborhoods and contribute to their economic vitality even though they aren't deliberately neighborhood oriented.

A DEVELOPMENT BANK

A unique arrangement has developed in the Chicago's South Shore, a community of 68,000 people which changed from virtually all white in 1960 to 95 percent black by 1974. During this period considerable disinvestment and deterioration occurred. In 1972 the South Shore Bank, the only commercial bank in the community, announced its intent to move out of the neighborhood, but after residents protested, the U.S. Comptroller of Currency, the regulator of national banks, denied the application. Instead, in 1973 a group of investors formed a bank holding company, the Illinois Neighborhood Development Corporation (INDC), and bought the bank with the intent of drawing in new deposits and making investments in the community. After getting the bank in better financial condition, INDC formed three subsidiaries: City Lands Corporation, a for-profit corporation to initiate and manage real estate rehabilitation projects; the Neighborhood Fund, a minority enterprise small business investment corporation (MESBIC), licensed by the Small Business Administration (SBA); and the Neighborhood Institute, a private nonprofit corporation, to initiate housing, social, and economic development projects. The bank's staff also organized the South Shore Development Company, a separate, SBA-licensed local development corporation (LDC) which makes below-market rate loans for new businesses and for expansion.

After eight years experience, an outside evaluator concluded that INDC and its subsidiaries had succeeded in demonstrating the following (Woodstock Institute, 1982:i):

- There are sound banking risks and good customers to be found among the businesses and property owners in lower income, predominately minority, urban neighborhoods;
- A bank can operate profitably in serving those customers without undue risk or loss;
- A bank holding company can operate community development subsidiaries to undertake riskier endeavors, serve lower income residents and complement the bank's efforts without impairing its basic banking business; and
- The combined commitment of the holding company and its subsidiaries can make a visible impact on neighborhood problems.

LAND TRUST

In some localities neighborhood land trusts have formed to acquire, hold, and develop vacant land. They organize as private, nonprofit, tax-exempt corporations. A land trust might acquire privately owned land through a straight cash sale, a bargain sale in which the land is partly donated and partly sold, or donation of the full value. Banks, developers, and other private owners gain tax benefits through such donations. A land trust might also acquire publicly owned land, such as vacant lots taken over by city or county because of tax delinquency, surplus pieces of highway right-of-way and other condemned land, and land in possession of federal agencies because of mortgage defaults and other reasons. The land trust then develops the site as a neighborhood park, community garden, or other open-land use. The trust can also build housing or some other kind of development or sell it to a private developer or a nonprofit corporation for that purpose. Some land trusts have traded land parcels with private owners for the mutual advantage of both parties (Trust for Public Land, 1979 and 1980; Institute for Community Economics, (1982).

During the last ten years approximately 80 neighborhood land trusts have formed and acquired land in a dozen cities, including New York, Newark, Boston, Atlanta, Cincinnati, and Oakland.

COOPERATIVES

A cooperative is a third major kind of economic enterprise found in neighborhoods (along with private and nonprofit). It is owned and controlled by its members, usually through a board of directors. In selecting the board and making other membership decisions, each member has one vote regardless of financial investment. Benefits are shared by members, such as lower costs, patronage rebates, and production dividends. Housing coops and food coops are the largest and most numerous varieties. Some food coops started as buying clubs run by volunteers, expanded enough to hire employees, but have remained a fairly small-scale operation. Other food coops are much larger, draw membership and other customers from a larger area, and might be part of a chain. The cooperative mode is also used to operate schools, medical service, prepaid legal assistance, automobile repair, and optical services.

There are also producer cooperatives. Most common are dairy coops and other rural enterprises. In recent years persons in urban areas searching for alternative economic models have formed small manufacturing cooperatives, sometimes in a neighborhood setting. The workers and managers are the principal members.

There has been interest in other forms of worker-controlled enterprise, such as partnership, employee-owned corporation (with marketable or nonmarketable shares), and nonprofit corporation. According to Neil G. Kotler (1978:6), "Most worker-controlled neighborhood enterprises are nonprofit associations or corporations, governed by boards of directors composed of full-time workers, or else mixed boards of workers and community residents. A variation of this is a board of full-time workers who make the day-to-day decisions and a second board of workers and community residents who make general policy."

COMMUNITY DEVELOPMENT CREDIT UNIONS

The credit union is a special kind of cooperative. As a voluntary association of members sharing a common bond, a credit union enables its members to save and earn interest and to borrow at relatively low cost. The common bond might be as employees or as members of a church, lodge, or trade association. In the 1960s the U.S. Office of Economic Opportunity encouraged and supported the formation of credit unions in poor neighborhoods. In 1972 the National Credit Union Administration, which regulates federally chartered credit unions, formally established the "urban residential" common bond as a new basis for forming a credit union (NCUEA, 1979a:1-2; also see Caftel, 1978).

Many residential-based credit unions function as community development credit unions. That is, they seek to use their resources to contribute to economic development of their neighborhood. As a resident-controlled institution, they help stem the outflow of capital and direct it toward community improvements.

Employment and Training

Most neighborhood economic development programs have job creation as a stated purpose, along with other activities. In addition, some programs focus specifically upon employment through means of training, job placement, and public service employment.

HISTORY

Prior to 1960 the U.S. Employment Service and its state affiliates handled most public-sponsored job services, state and local vocational institutions were in charge of classroom occupational training, and a network of apprenticeship councils took care of that approach to on-the-job training. This changed rapidly during the 1960s as the Manpower Development and Training Act of 1962 and the Economic Opportunity Act of 1964 spawned numerous neighborhood- and community-based employment and training organizations. Most visible were three national networks: Opportunities Industrialization Centers (OICs), initiated by Rev. Leon Sullivan in Philadelphia; local affiliates of the National Urban League, which got into on-the-job training and apprenticeship programs; and units of SER-Jobs for Progress, which originated in the Southwest and expanded to midwestern and eastern cities. The first two were oriented toward black communities, the third toward Hispanics. They became known as community-base organizations (CBOs) and tended to concentrate in black ghettoes and Hispanic barrios, thereby becoming important neighborhood institutions. Community action agencies got into employment and training and so did numerous independent, neighborhood-based organizations (NBOs) in black, Hispanic and white ethnic neighborhoods. They continued during the 1970s, though came under the funding control of local prime sponsors, based in local government, under the Comprehensive Employment and Training Act (CETA) of 1973. This system was altered by the Job Training Partnership Act of 1982, which placed private industry councils in charge of local delivery systems, but CBOs were expected to continue operation of employment related activities. (For more on this history, see Levitan and Mangum, 1969; Snedeker and Snedeker, 1978; Anderson, 1976; and Hallman, 1980.)

TRAINING AND PLACEMENT

Employment and training activities can be viewed as a system of sequential steps through which a person moves from being unemployed to working regularly. The steps include recruitment, interviewing and testing, counseling and decisions on next steps, class or shop training, sheltered work experience, on-the-job training, job placement, supportive services. Not everyone needs all of these services, and persons start at different stages. But they do need continuity.

Community action agencies sponsored neighborhood employment centers to offer as complete services as possible. These usually housed their own staff plus personnel from the state employment service, vocational rehabilitation agency, and other organizations. OICs, Urban League operations, and SER units had a heavy focus on training and placement, but they also recruited and counseled. Smaller neighborhood organizations tended to take on fewer components.

Federally sponsored experiments leading up to CETA brought local government more fully into employment and training, and after passage of the 1973 act a number of cities took charge of neighborhood employment centers, using them as decentralized units of their larger operations. Nevertheless, much of the basic recruitment, counseling, and placement activities have continued to be neighborhood based. So is part of the training, though vocational schools tend to serve a number of neighborhoods or the whole city. Because jobs are located throughout the metropolis, the most successful programs have effective linkages between neighborhoods, the various agencies offering training, and employers (Hallman, 1980; Urban Systems Research and Engineering, 1979).

PUBLIC SERVICE EMPLOYMENT

Quite a few neighborhood organizations were involved with the Neighborhood Youth Corps during the 1960s, and also with a new youth employment initiative in the latter part of the '70s. As a part of these programs youthful workers and trainees engaged in neighborhood improvement activities as well as receiving job training and counseling.

Neighborhood organizations also became involved in public service employment (PSE) activities under the 1973 act, expanded in 1976. At its peak in 1979, about one-third of the CETA PSE participants were assigned to nonprofit organizations, including citywide social service agency and neighborhood-based organizations. The latter handled an estimated ten percent of the total PSE jobs. They undertook a wide range of projects: housing rehabilitation, weatherization, cleanup and beautification, food programs and services to the aging, arts and cultural activities, day care and family services. A number of the housing improvement and service delivery programs we reviewed in the previous three chapters made productive use of CETA workers. A national evaluation concluded, "On balance, it appears that although workers are seen as relatively unskilled, the services they provide are important and likely to be appreciated by a substantial segment of the communities from which they are drawn or for which they provide services" (Cook et al., 1981:140; also see Nathan et al., 1981; Mirengoff et al., 1979 and 1982; and Urban Systems Research and Engineering, 1981).

Analysis

Most of the economic development programs we've considered have been carried out in low and moderate income neighborhoods. Although not totally bereft of private enterprises, these neighborhoods have felt the need for organizing new community institutions to tackle their economic problems and draw in new resources. In some places, such as the Northeast and industrial Midwest, they have had to work in an atmosphere of a weakening metropolitan economy. Everywhere they've had to live with the ups and downs of the national economy and shifting federal policies on economic development programs. It's not been easy, and our emphasis on accomplishments shouldn't shield the difficulties and the inevitable failures of some neighborhood-based economic enterprises.

Although these neighborhood economic development organizations rarely articulate a well-developed micro economic theory, they function in the context of a little economy linked to a broader economy. They are particularly concerned with the flow and use of wealth and money, and they have a concern for enhancement of human resources.

FLOW OF WEALTH AND CASH

As we studied Chapter 6, all neighborhoods experience a flow of money and wealth: into, out of, and within. They differ as to their basic wealth, the total and per capital income of residents, how much flows in and out, how fast money circulates within the neighborhood before leaving, and whether on the whole the neighborhood has a net gain or loss.

A number of the programs reviewed in this chapter aim at increasing internal circulation and stemming the outflow. That is one purpose of community development credit unions and the South Shore Bank. Commercial revitalization programs want residents to do more of their shopping in the neighborhood. So do the retail enterprises sponsored by community development corporations and minority enterprise development programs. Housing reinvestment programs, as described in Chapter 14, are based partly on the idea of converting the residents' savings into investments within their own neighborhood.

A number of the programs also aim at bringing more capital and more consumer spending into the neighborhood. Capital infusion is a major thrust of community development corporations, and the success of commercial revitalization depends upon attracting outside investments. New manufacturing enterprises sponsored by CDCs try to sell to a market beyond the neighborhood. Revitalized neighborhood shopping areas seek customers from as wide an area as possible. Neighborhood banks and credit unions try to use their investments to leverage funds from outside sources. (For further reading, see articles in Friedman and Schweke, 1981.)

As employment and training programs make residents more employable and get them jobs outside the neighborhood, they have the effect of increasing the inward cash flow of personal income.

HUMAN RESOURCE DEVELOPMENT

These latter programs, though, aim first of all at helping residents develop their full capacity to participate in the world of work, wherever the jobs are. They provide new job skills, teach the inexperienced how to get a job, and offer supportive services where necessary. Day-care, health services, and transportation programs held remove obstacles which keep some persons out of work.

The jobs created by expanded businesses, nonprofit organizations, and governmental agencies and by public service employment programs tap the unused resource of the unemployed.

Technical assistance provided by CDCs and commercial revitalization projects help merchants, service providers, manufacturers, and other entrepreneurs to increase their managerial and technical capacity. These programs also help individuals start new businesses and assist small entrepreneurs to expand. The CDCs themselves have needed and benefit from outside technical assistance.

NEIGHBORHOOD CONTROL

These varied economic development activities also enhance neighborhood control of elements of the local economy. This is achieved through establishment of new institutions, such as community development corporations, housing development organizations, merchants associations, commercial revitalization task forces, credit unions, cooperatives, land trusts, and development banks. Although they usually don't have political governance as their aim, they run parallel with government-oriented efforts of neighborhood control. Indeed, as we'll see in the next chapter, the experience of neighborhood corporations (with which CDCs are one variety) offers useful insights on the feasibility of more complete neighborhood self-governance.

Neighborhood economic development is an arduous process, fraught with difficulties, and constantly facing uncertainties of market forces. Yet looking back at 20 years of experience, we can see considerable progress and many successes, along with the setbacks and failures which seem to be the inevitable part of any learning experience. What some people and organizations in some neighborhoods have done, others could likewise do.

Exercises

(1) Locate and study a neighborhood economic development operation of some variety.
(2) Find out its origins, who started it, whether it evolved out of some other thrust (such as protest), and whether its leadership has changed.
(3) Describe its current programs and sources of funds. Analyze its accomplishments and failures and determine the reasons. Construct a balance sheet.
(4) What view does this operation have of economic dynamics within the neighborhood and relationships to the broader economy? What aspect of neighborhood economics does it emphasize?
(5) Would you be willing to invest in a neighborhood enterprise as a resident? As a nonresident? Which type? Why, or why not?

Note

1. This section draws upon the following sources (abbreviated in the test): Goldstein and Davis, 1977; National Center for Urban Ethnic Affairs (NCUEA), 1979B; National Council for Urban Economic Development (NCUED), 1979; National Development Council (NDC), 1979; National Trust for Historic Preservation, 1979; and National Urban Development Services Corporation (NUDSC), 1977.

CHAPTER 16

GOVERNANCE

When we examined the concept of neighborhood as a political community in Chapter 5, we noted that it has three major dimensions: the neighborhood acting politically in its dealings with the city, county, state, other governmental units, and private sector interests; broader governmental domains conducting operations within the neighborhood through administrative decentralization; and the neighborhood taking responsibility for particular activities through instruments of self-governance. Regarding the latter, we noted that it actually occurs in suburban units of neighborhood size, in small enclave cities, and in special districts in unincorporated suburban areas; that central city neighborhoods have formed neighborhood councils and neighborhood corporations which have quasi-governmental characteristics; and that a number of persons have proposed full-fledged neighborhood government for city neighborhoods. In this chapter we concentrate upon experience with resident-controlled governance during the last 20 years.

Suburban Experience

The suburbs of most metropolitan areas within the United States have numerous units of local general government, such as municipalities, towns, townships, and villages, and also counties at a larger scale. The pattern varies among the states.

Some states have completely divided their territory into cities and towns or townships. In most of these states when a new municipality incorporates, its territory is withdrawn from township government. However, in New York and six other states, towns can have small incorporated villages in their midst, and in Indiana townships overlap the municipalities. Towns or townships have virtually the same powers as cities in the New England states, New York, New Jersey, Pennsylvania, Michigan, and Wisconsin. With some exceptions, townships are weaker in Ohio, Indiana, Illinois, Minnesota, North Dakota, South Dakota, and Kansas, but they do provide some municipal services to suburban residents. Of the other states with townships, in Iowa they have atrophied to minor subunits of county government, Missouri and

Nebraska have eliminated them from metropolitan areas, and they exist in Washington only in Spokane County.

In other states when a residential subdivision is developed outside city limits, municipal services are provided in several ways: by the county, by incorporating a new municipality, through one or more special districts, or a combination. Incorporation gives residents their own general purpose government, perhaps shared with other subdivisions. They can then choose to run services themselves or to contract with the county or a private company for specific services. As an alternative, the inhabitants can form special districts to handle particular functions, such as fire protection, water supply, sewer lines, street lighting, and libraries.

In 1980, 40 percent of the metropolitan population lived in central cities, 41 percent resided in suburban municipalities or strong towns or townships, and 19 percent lived in unincorporated areas or weak townships. Some of the suburban jurisdictions were large, but many were of neighborhood size or encompassed several neighborhoods with similar characteristics. Altogether 23 percent of the metropolitan population lived within the territory of strong general purpose governments with population under 25,000 (U.S. Bureau of the Census, 1983). Thus, in a very real sense nearly one quarter of the metropolitan population now has neighborhood government.

POWERS OF GENERAL PURPOSE UNITS

To illustrate the kind of powers small general purpose units can possess, let us take the example of six small cities which were once suburban municipalities but are now completely surrounded by a central city (Hallman, 1974:45-49). These enclave cities are St. Bernard and Norwood encompassed by Cincinnati, Bexley and Whitehall within the boundaries of Columbus, Ohio, and Highland Park and Hamtramck in the midst of Detroit. They are all mayor-council cities, and all but one have a full-time mayor. Two have city councils chosen at large, and the other four combine district and at large election. Council size ranges from five to nine. Elections are nonpartisan.

Most of their employees and local funds are concentrated on three functions: police, fire protection, and public works (including sanitation, streets, and maintenance of public facilities). Each city has its own police force, which handles routine patrol, traffic control, and elementary crime investigation, and each has its own police communications system. They all have mutual assistance agreements with the larger city, and four of them rely on an outside agency for police training. Five of the cities have their own fire departments, but with reciprocal agreements with surrounding jurisdictions. The sixth contracts with the central city for fire protection.

All six cities collect refuse. Two of them handle their own disposal while the others rely upon private or regional landfill sites. Each city cleans and repairs streets, removes snow, handles street lighting, takes care of street trees, and keeps sidewalks in repair. One city has a complete water system, three of them buy water wholesale from the central city, and residents of the other two are served directly by the central city. Each city builds and maintains local sewer lines but ties into trunk lines connected to treatment plants of larger jurisdictions (central city or regional). Four of the cities have small units for sanitary health inspection, but the other two rely upon the county health department.

All six cities own parks and administer recreation programs. Two have their own libraries, and four have branches of the county library system. Independent school districts handle public education. Each city has a zoning ordinance. Four of them have participated in the federal community development program, and three have received federal employment and training funds. The two Michigan cities have municipal courts with elected judges. The four in Ohio have a mayor's court for traffic violations, disturbing the peace, and other misdemeanors under city ordinances. Each city has a small jail, mostly for temporary custody of offenders.

With population ranging from 5,400 to 28,000, these six enclave cities are like functional neighborhoods, or community districts of several neighborhoods, within the central city. They have a revenue base of property taxes, shared state taxes, state and federal grants. They meet three major tests for participation in a federal system: independence within its ambit of power, financial means to handle its responsibilities, and political ability to deal with other levels.

LOCAL SERVICE DISTRICTS

Special districts come in many sizes and do many things, ranging from local services to regional transportation. Small independent districts with property taxing power provide residents in unincorporated suburbs a measure of self-governance. Of this variety three-fourths are found in seven states: Washington, Oregon, California, Colorado, Nebraska, Illinois, and New York. The first five lack townships; in Illinois township government is weak though stronger in New York.

Another dozen states, mostly without townships, have most of the remainder of special districts with property taxing authority.

In reviewing experience of such special districts in California, Robert B. Hawkins, Jr. (1976:66) observed that "citizens served by districts generally demonstrate higher rates of satisfaction than do those served by larger and more unified local government structures." Hawkins concluded that this experience was relevant to inner city neighborhoods (116-117):

> If citizens have relatively attractive options to form districts, existing officials find themselves in a competitive local government environment. Their constituents can bargain and negotiate for the types and levels of service they desire. For example, if minorities in large cities had the option of forming neighborhood governments with independent powers to tailor the provision of goods and services coming into their community, it is highly likely that elected officials and managers would be more responsive to their demands; should officials ignore their demands the voters could have the option of forming such a unit, thereby reducing official authority.

Michael Silver (1981) has also indicated that creation of special districts offers neighborhoods an opportunity to have their own government.

Within The Cental City

Although residents of central city neighborhoods haven't had the option of forming their own special districts with taxing authority, they have found other devices to use in getting things done for themselves. We'll look at two types: neighborhood corporations and neighborhood councils. Neither is a fully developed government in legal form, but in a broader sense they constitute a means of self-governance. Furthermore, their experience provides a basis for assessing whether fuller neighborhood government is feasible and desirable.

NEIGHBORHOOD CORPORATIONS

As we have seen in the previous three chapters, many neighborhoods have formed nonprofit corporations during the last 20 years in order to carry out activities of physical and economic development and to deliver particular services to residents. In many respects neighborhood corporations are a result of the human rights movement which has stressed that people should have greater control over programs and services affecting their lives.

Rather than attempting to take over existing services, most neighborhood corporations have moved into program areas where new funding was becoming available, especially federal grants. At first many got direct federal support under categorical programs, but as block grants were established, first for community development and employment programs and more recently for community and social services, they have had to deal more with local government. From the beginning some of them have obtained foundation money and, to a lesser extent, private corporate donations. A small number of neighborhood corporations have engaged in profit-making enterprises (sometimes through a subsidiary). Federal funding cutbacks have pushed more of them to seek alternative sources of support.

Here are some of the principal varieties and their initial funding sources, many of which we've examined in previous chapters:

- *Neighborhood boards and community corporations* initiated under the Community Action Program, both locally and through national demonstrations; similar bodies later funded as part of Model Cities Program.
- *Program corporations,* such as Head Start, neighborhood health centers, multiservice centers, supported by Community Action and Model Cities funds.
- *Community development corporations (CDCs),* emphasizing economic and community development, set up under an amendment of the Economic Opportunity Act and some of them supported by the Ford Foundation.
- *Community-based organizations* (CBOs) instigated as demonstration projects under the Manpower Development and Training Act of 1962, as amended, and later folded into the Comprehensive Employment and Training Act of 1973.
- *Neighborhood development organizations* (NDOs) with a housing and physical development emphasis, evolving from Model Cities and foundation projects, expanded under the Community Development Block Grant Program, and some of the receiving money from the Neighborhood Self-Help Development Act of 1978.
- *Neighborhood housing services* (NHS), governed by a tripartite board of residents (majority), lenders, and local government officials; funding and technical assistance from Neighborhood Reinvestment Corporation (a federal agency) and also local support.

- *Youth program operations,* at one time getting funds as juvenile delinquency prevention projects; more recently garnering local support, making use of volunteers, and tapping a variety of federal programs.
- *Community crime prevention programs,* supported by the Law Enforcement Assistance Act; some of these organizations also involved in neighborhood development.
- *Tenant management corporations* in public housing.
- *Neighborhood organizations contracting* with city government to run municipal services.

Because most neighborhood corporations derive the bulk of their funds from outside their neighborhoods, they inevitably act as interest groups in dealing with local government and federal and state agencies because these sources operate in the political arena. Beyond their immediate financial self-interest, many neighborhood corporations are committed to overall improvement of the lower income areas where they function and to gaining benefits for the residents. This orientation draws them into advocacy. From a tactical viewpoint, their capacity, or potentiality, to mount protest and use other advocacy techniques increases the likelihood that governmental funding sources will respond to their financial needs.

NEIGHBORHOOD COUNCILS

Whereas neighborhood corporations control operations but lack an official connection with city government, neighborhood councils are the converse, having official recognition but generally not engaging in direct operations. (See Chapter 5, p. 70 for a fuller definition of neighborhood council.)

Dayton's seven neighborhood priority boards, for example, receive information about and make recommendations on planning and zoning decisions affecting their neighborhoods, liquor license applications, proposed land sales, and appointments to city boards and agencies. Their membership ranges from 26 to 45. Most members are elected from geographic subareas, but two boards have some at-large members and one board, covering a large district, has representatives from smaller neighborhood associations. The board in the downtown area has members from housing complexes, the police district, churches, and other downtown organizations, which choose their own representatives at regular meetings. The other six boards rely upon mail ballots for the election process.

A major activity of Dayton's priority boards is the annual preparation of need statements. These are developed through public hearings, neighborhood surveys, discussions with citizen associations and priority board committees, and analysis of complaint records at the city's site office for their area. Each board ranks its needs. The Neighborhood Affairs Division, which provides staff support through the site office, assembles all the need statements and submits them to the city manager, who transmits them to city departments for review and comment. Departmental responses are transmitted back to the priority boards. This process provides input into the city's operating and capital budgets, the community development (CDBG) program, and the city's state and federal legislative program. In 1982 the city asked all priority board members to respond to a questionnaire evaluating city services and neighborhood conditions. The board chairpersons get together regularly to sort out priorities among themselves and to consider common issues. They meet monthly with the city manager, serve on the CDBG advisory task force, meet from time to time with the City Commission, and also deal with other community institutions, such as the Board of Realtors and lenders.

Other cities, and a few counties, having neighborhood councils also use them as an official channel of communications from local government to the neighborhood on such matters as proposed zoning changes, liquor board license applications, street and park improvements, capital budget matters, vacancies on citywide boards, and CDBG planning (for low- and moderate-income neighborhoods). Some neighborhood councils develop neighborhood plans which become part of the official master plan, following review, possible review, possible modification, and approval by the planning commission and city council. Neighborhood councils also play an advocacy role, though they tend to work through official channels rather than to use protest tactics. Some of them sponsor self-help activities (Hallman, 1977a; also see Andersen, 1979; Woody, Walters, and Brown, 1980; Rosenbaum and Rich, 1982; Rohe and Gates, 1983).

The latest trend, evidenced in several cities, is to bring neighborhood councils into the local budgetary process at an early stage, even before departmental budgets are prepared (Hallman, 1980). The starting point for neighborhood councils might be an annual assessment of neighborhood needs, as occurs in Dayton. In Portland, Oregon, for instance, neighborhood associations prepare a series of individual need reports, describing the problem, indicating how many persons are concerned about it, suggesting solutions, and giving it a priority rating. In New York City the community boards (which serve multineighborhood community districts) prepare an annual statement of needs and priorities, which they submit along with their budget requests.

An alternative approach is for neighborhood councils to draw up neighborhood plans or work programs and make annual revisions. St. Paul's district councils prepare district plans dealing with physical, social, and economic conditions, and they propose long-range solutions for ameliorating problems and maximizing neighborhood assets. These plans form the basis for neighborhood input into the city's capital budget. Atlanta's neighborhood planning units produce annual plans describing projects which the neighborhood desires. Cincinnati's community councils develop annual work plans, indicating priority of needs they want the city to address.

In neighborhoods where the federal Community Development Block Grant (CDBG) Program is active, neighborhood councils use their needs assessment and planning processes as the basis for recommending priorities for CDBG finding. Indeed, Birmingham, Alabama set up its neighborhood structure primarily for the purpose of CDBG citizen participation, though neighborhoods get involved in other issues, too. There are 93 neighborhood advisory groups, whose three officers are elected bienially. Each selects a representative to a community advisory board encompassing several neighborhoods, and the presidents of these 21 boards constitute the citywide Citizens Advisory Board (CAB). Each neighborhood receives a small annual allocation of public improvement funds, derived from federal revenue sharing and local general revenue. Low- and moderate-income neighborhoods receive CDBG funds as well, and five neighborhood strategy areas at a time receive substantially more CDBG funds for a three year period. Neighborhood input is highly influential in the use of these funds. The Citizens Advisory Board convenes monthly to consider a variety of citywide issues, meets with the mayor and City Council at least quarterly, and from time to time invites city department heads to meet with its members and provide information. As in other cities with neighborhood councils, Birmingham's neighborhood advisory committees receive advance notice of prospective city actions affecting their area.

In all of these cities with neighborhood councils, final authority remains with city council in both the local budget and CDBG program, and with boards and departments on matters within their authority. Nevertheless, neighborhood councils often get what they want through a combination of persuasion and pressure. Having official recognition, an open process for determining neighborhood priorities, and access to public officials, they are in an excellent bargaining position in the wider policy process.

ANOTHER EXPERIENCE

In an interesting variation, the 44th Ward Assembly of Chicago functioned as a special kind of neighborhood body from 1972 until 1979. Alderman Dick Simpson created it and pledged that he would "be bound by the decisions of the 44th Ward Assembly on important issues before the City Council and on projects undertaken to promote the welfare of the citizens of this ward and this City, provided that those decisions are either unanimous or approved by a 2/3's vote as outlined in the Assembly's Charter." Membership consisted of two elected delegates from each precinct and delegates appointed by various community organizations. Meeting monthly, the assembly discussed community issues and pending city council legislation. It also undertook its own projects, such as an annual fair, food collections, and Operation Whistlestop for crime prevention. A parallel Asamblea Abierta formed to achieve greater involvement of the Hispanic population. These operations received staff support from the ward office and small outside grants. Simpson didn't run for reelection in 1979. His elected successor was in the process of organizing a new assembly when he resigned, and his replacement from the regular Democratic organization let the process lapse (Simpson et al., 1979; Salem, 1980).

Evaluation of Experience

In sorting out this highly diverse experience, we can offer a number of findings. First, it is indeed possible to operate a wide variety of public services at neighborhood scale. Thousands of suburban municipalities, towns, and townships are doing so effectively, and numerous neighborhood corporations are, too. Second, although cost comparisons aren't available, it appears that unit costs of neighborhood-size operations compare favorably with larger operations for many services. Labor-intensive services are particularly suitable for small-scale administration while capital-intensive projects tend to require a larger base. Third, neighborhood-based operations can't handle all the service needs of residents so that neighborhood activities have to be part of larger systems, using such coordinating devices as we reviewed in Chapter 12.

PROGRAM ACHIEVEMENTS

Chapters 12, 13, 14, and 15 have laid out some of the program achievements of neighborhood organizations. We have also previously noted the finding by Robert K. Yin and Douglas Yates, following a review of 215 case studies, that strong forms of decentralization, that is, where clients had direct governing control over a service, were statistically associated with higher frequencies of improved services as compared to weaker forms (1974:59-68). A study by Richard L. Cole encompassing 26 neighborhood programs in six midwestern localities found that neighborhood involvement is likely to "achieve at least in the participants' own judgment, a more favorable allocation of goods and services (material awards)" (1974:134). However, he noted that programs with moderate rankings in intensity and a scope of citizen participation were considered most successful by participants, rather than those with least or most participation. In examining seven different approaches in New York and New Haven, Douglas Yates concluded that the amount of impact achieved wasn't related to the amount of power vested in the experiment but rather whether it focused upon concrete service problems rather than broadly defined urban problems (1973:66).

The Task Force on Governance, Citizen Participation and Neighborhood Empowerment of the National Commission on Neighborhoods sponsored case studies of 44 neighborhood organizations spread around the nation and judged that these organizations had successfully "taken on a wide variety of complicated tasks which have resulted in physical revitalization, improved service delivery, and innumerable events which have developed a sense of future for the community" (1979:App. I, 34).

Urban Systems Research and Engineering, Inc. examined the performance of community-based organizations (CBOs) working under 22 local prime sponsors under the Comprehensive Employment and Training Act (CETA) and concluded (1979:102):

(1) In many Prime Sponsors, CBOs have been more successful in attracting and serving minorities and other disadvantaged men and women than have any other alternative service deliverers.

(2) No patterns have emerged concerning the relative effectiveness of CBOs and other service deliverers. CBOs have out performed non-CBOs in some instances and have performed more poorly than non-CBOs in others.

A Brookings Institution study of public service employment program (PSE) in 42 local jurisdictions discovered that 99 percent of the PSE jobs with nonprofit organizations in the basic PSE program were new jobs compared to 84 percent for school districts and 76 percent for local governments. The remaining jobs constituted substitution of federal for local dollars to fill existing positions (Nathan, 1978:34).

In its study of 99 neighborhood development organizations, the Urban Institute identified the following concrete results (Marshall and Mayer, 1983:9-10).

> Finished and unfinished projects together have already generated over 1800 renovated housing units, over 200 weatherized homes and solar installations, more than 125 new housing units (and many more under construction), four community facilities, nearly 80 permanent jobs in economic development efforts (neglecting jobs created as part of housing, commercial revitalization, etc. work), over 80,000 square feet of renovated or newly built commercial space, 11 new NDO-owned ventures and assistance to almost 50 businesses.

This condensed sample of evidence offers some sense for achievements of neighborhood organizations. One comes away from reading these and other studies with a feeling that many neighborhood organizations are demonstrating a capacity to carry out programs effectively, others have failed, and some are in the middling state of so-so performance. The bell-shaped curve, which statisticians apply to other phenomena, seems to describe the program achievements of neighborhood-controlled operations.

EFFECT UPON RESIDENTS

In addition to their program achievements, neighborhood operations have other positive effects upon the residents.

Programs creating new roles and new participation opportunities for neighborhood residents have greatly contributed to leadership development in poor neighborhoods. This was the greatest latent effect of the Community Action and Model Cities programs, and it continues to occur in other programs. Thus, the National Commission on Neighborhoods found the most striking feature of its case studies to be "the growth and development of literally thousands of new neighborhood leaders, accepting new responsibilities in their neighborhoods and for their neighborhoods' improvement" (1979:App. I, 59). The commission indicated that this had personal impact on the participants: a new sense of dignity and self worth and the acquisition of new skills leading to employment opportunities. Numerous individuals from two decades of neighborhood programs have moved to leadership positions beyond their neighborhoods.

This reflects the way participation has enhanced residents' political efficacy, that is, their sense that they can be effective in politics. This was one of the positive results noted by Richard L. Cole in this 1972-1973 survey. It is also consistent with broader studies of participation in America by Sidney Verba and Norman H. Nie, who have shown that individuals participating actively in community organizations and political parties have a much stronger feeling of political efficacy than people who limit their activity to voting, only to pursue individual interests, or are inactive (1972:85-89).

Another effect is changes in residents' attitudes. Cole found that programs of neighborhood involvement are likely to "increase the participants' confidence and trust in local officials." He presented similar findings in a later study of four neighborhood health and mental health centers, reporting that "each respondent was asked if participation had any impact on mutual trust and respect among clients, trustees, and staff personnel. Overwhelmingly, the responses were 'definitely yes'" (1981:56).

EFFECT ON BUREAUCRACIES AND OTHER INTERESTS

Increased citizen participation and neighborhood decentralization have sometimes had adverse effects upon agency personnel and other special interests. Thus, Yin and Yates indicated that improved agency attitudes occurred least out of five outcomes they surveyed, occurring in only about one out of six cases reporting (1974:62-63). David J. O'Brien has explained that the organizational maintenance need of public bureaucracies, particular in an era of fiscal crisis, is to keep services to a minimum even though residents are demanding increased services (1975:9-10). And Barry Checkoway and John Van Til point to the emphasis of administrative agencies on efficiency, economy, and control and their perception that participation is likely "to cause long delays in action, to expand the number and intensity of conflicts, and to increase the cost of operations" (1978:33). They are talking about perceptions, whether they are accurate or not (also see Kweit and Keil, 1980).

The underlying factor is power. This was most vividly dramatized by the Ocean Hill-Brownsville dispute in New York in the late 1960s which pitted community activists against the powerful teachers union and the central school administration. It was present in contests for control over the Community Action Program, Model Cities, more recently the CDBG program, and varied other local programs and services. Nevertheless, in each of these programs numerous examples exist showing how citizens, administrators, and governing officials worked out their differences. And in recent years neighborhood organizations have gained greater skills and political influence, and elected officials and administrators in many localities have been more willing to recognize the legitimacy of neighborhood action. Yet, it has often taken protest tactics to awaken the bureaucracies of the need to change (Lamb, 1975; Cunningham and M. Kotler, 1983).

REPRESENTATIVENESS

Another issue, particularly applicable to neighborhood councils, is how representative neighborhood governing bodies are of the neighborhood population.

During the early days of the Community Action Program, there was some indication that those chosen to be on the community action board were not as poor, were better educated, and had other characteristics different than the average poor person they were supposed to represent (Bowen and Masotti, 1968; Pollinger and Pollinger, 1972). However, in studying boards of community development corporations, Rita Mae Kelly determined that, while better off than the average resident, "the composition of the governing boards still reflects the communities to the extent that most politically constituted governing boards in the United States, particularly nonpartisan ones, do" (1977:72).

Reporting on studies of elected neighborhood councils in New York, Raleigh, St. Paul, and Washington, D.C., Richard C. Rich (1983) has indicated that the members "tend to be among the better educated, more affluent and more professional citizens of their communities," but that "the opinions and perceptions of the leaders (both officers and activists) tend to parallel those of the public in their neighborhoods." Charles White and Sheldon Ender have made similar findings for Portland, Oregon, where they compared persons who were active in officially recognized neighborhood associations with the general population. They found that "the characteristics of those who do participate tend to reflect those of the neighborhood within which these organizations function." Furthermore, "the participants in neighborhood associations generally express perceptions of neighborhood conditions which are similar to those of nonparticipants." Also, participants and nonparticipants rate city programs similarly (1981:51-52).

As with other representative bodies in American democracy, the system can be open to full participation, can encourage and facilitate widespread involvement, but it is ultimately up to the electorate to assure that they are accurately represented.

SUMMATION

If we look at the movement toward neighborhood control developmentally, we can see that enormous strides have been made during the last 20 years. Compared to the initial neighborhood corporations of the mid-1960s groping their way into unknown territory, today there are hundreds of quite sophisticated operations. Compared to the harshness of the power struggles of the 1960s, numerous neighborhood organizations, public officials, and other involved interests have learned how to moderate their demands and responses and to work out compromises reasonably satisfactory to most of the contending parties. Neighborhood councils are functioning harmoniously in many cities. Neighborhood corporations are achieving results approximately on par with public agencies and citywide voluntary organizations, that is, some excellent, some poor, most in between in the generally satisfactory range. Although neighborhood organizations don't have the answer to some of the broader societal issues, such as how to achieve full employment, they are showing that they can contribute to community betterment within their sphere of operation. They have proven that self-governance is possible in city neighborhoods. In the process they strengthen the social fabric and enhance the sense of community.

We'll build upon this experience in the next chapter when we examine more fully how neighborhood government within central cities might be shaped.

Exercises

(1) For a small, incorporated suburb, identify the activities handled by the municipality, town, or township. Identify public services handled by other jurisdictions.

(2) Study a neighborhood corporation: when, how, and by whom it was organized; what it does and its accomplishments; where it gets its funds; its relationships with residents, local government, and the private enterprise sector.

(3) Study a neighborhood council, or if none in your locality, a neighborhood organization or advisory committee with good connections or quasi-official recognition by the city or county. When, how, and by whom was it organized? What are its activities and accomplishments? Does it have staff support? If so, what kind, whose auspices, and how funded? What is is relationships with residents, local government, and the private enterprise sector?

(4) Compare the neighborhood corporation and the neighborhood council as to roles, accomplishments, and relationships. Are there circumstances in which one form is superior to the other?
(5) Make a list of arguments favoring and opposing greater neighborhood self-governance in the central city.
(6) Do you favor neighborhood government? Why or why not?

PART IV

STRATEGIES FOR THE FUTURE

CHAPTER 17

ACHIEVING NEIGHBORHOOD WHOLENESS

We have covered a broad expanse in considering the place of neighborhoods in American life. We have examined the neighborhood as a personal arena, a social community, a physical place, a political community, and a little economy, and we've explored how these manifestations relate to the wider city and metropolis. We've traced the evolution of neighborhood as a concern of individuals and organizations for a century and have studied techniques used by community organizers. We have looked at a broad range of neighborhood action: self-help activities, decentralized municipal services, human services, public education, physical preservation and revitalization, economic development, and governance. In this part we bring the book to completion by considering some strategies for the future, especially how to achieve neighborhood wholeness and how outsiders can be partners and supporters of neighborhoods.

CONCEPT OF WHOLENESS

We should strive to achieve neighborhood wholeness. By wholeness I am referring to several meanings of the word: *complete,* that is, all essential parts in place and working properly; *functionally integrated,* so that various parts reinforce each other; and *healthy,* with wounds healed, illnesses cured, and wellness maintained. However, I'm not advocating solitary and exclusive action, separated from the broader community, nor am I looking for perfection.

The neighborhood in its relationship to the broader community bears some resemblance to de Tocqueville's description of American federalism as consisting of "two distinct social structures, connected, and as it were, encased, one within the other" (n.d.; book I, 60). A more dynamic characterization would be a small network which is part of and connected to a larger network. Or, a set of small networks within the neighborhood connected to sets of larger ones. For, neighborhood wholeness can't be achieved in total isolation.

Immediate Personal Neighborhoods

Each of us as an individual strives to achieve wholeness in our own life. While ultimately we have to put the pieces together for ourselves, we can gain wholeness only through association with other people. This starts within our household and extends outward, starting with our immediate personal neighborhood. As we saw in Chapter 2, this may consist of contact with people living next door, across the street, on the same courtyard, or on the same floor of an apartment building. This is where neighboring occurs, that is, informal social relationships and mutual aid.

Neighboring can be facilitated through design features which bring people together naturally and don't drive them into isolation within their quarters. Security measures also help so that people aren't afraid to be out in their yards, on the street, in courtyards, and in other public spaces. Conversely cooperative efforts to assure security, such as block watches and neighborhood walks, strengthen neighborly feelings. Social events, joint projects for beautification and cleanup, and organized exchanges (such as babysitting cooperatives and other forms of bartering) also foster neighborliness.

Sometimes these things happen quite informally, but many neighborhoods have found that conscious block organizing is exceedingly useful. Sometimes residents living on the same block have never met and have welcomed the opportunity to meet their neighbors. There are, though, people who prefer to be left alone, and should be if that's their choice, but not as many as you would imagine. (For a review of block organizing techniques, turn again to Chapter 10.)

It is easier to achieve rapport among residents within an immediate personal neighborhood where they have similarities. But this isn't an argument for absolute homogeneity, especially in race, ethnicity, and social class, for there are other similarities which bond people, if allowed to surface. For example, parental concern for their children cuts across racial and class lines. Parents have similar interests in their children being safe, staying out of trouble, and getting a good education, even though they might define differently how these objectives can be best achieved. By starting where they agree and supporting one another there, residents can more easily accept and cope with differences.

Encouraging residents to act as caring neighbors is another way to strengthen block life. We learned in Chapter 3 that every neighborhood has natural helpers to whom others turn, in Chapter 11 we came across self-help activities using volunteers to help their neighbors, and in Chapter 13 we saw how it was tied into broader approaches of human services. Caring neighbors can help fellow residents cope with crises, uncover and lessen child abuse, restore juvenile delinquents and released adult offenders to productive social roles, and strengthen the will of reformed drug users. Although training and orchestration might come from an organization serving the entire neighborhood, many acts of caring neighbors are performed at the intimate scale of the immediate personal neighborhood.

Functional Neighborhoods

Much of our effort to achieve neighborhood wholeness should be focused upon what I have referred to as the functional neighborhood. This is the territory typically encompassed by a neighborhood association and defined by city officials for planning purposes. It is large enough to provide an effective base for activities and services responsive to basic human needs, such as the area served by an elementary school, a parish church, local shops, a recreation center, library, police patrol, social service operations, and political party units. Although these service areas may overlap, compromises can be worked out to define distinct neighborhood boundaries for organizational purposes.

COMPLETENESS

Residents of functional neighborhoods should be assured that a complete array of needed services are conveniently available to them. This doesn't mean that everything must be based within the neighborhood or that every service must be under neighborhood control. Rather it provides a test for evaluating the existing constellation of services and service deliverers for their contribution to neighborhood wholeness.

By way of illustration, Table 17.1 lists 18 human needs of urban dwellers which are met at least partially within neighborhoods. For any neighborhood it is possible to go through this list and indicate how, by whom, and where these needs are met, or not being met adequately. In many instances the response is a combination of individual and family effort, informal groups, voluntary agencies, private enterprise, and government.

We can take each function listed on this chart, analyze the needs of each neighborhood, determine how well the array of services and private opportunities within and outside the neighborhood meets those needs, and design activities and connections to respond to unmet needs. As we do, we'll find that completeness means different things in different neighborhoods.

Take employment, for instance. Some residents may be self-employed and work at home, others have jobs within the neighborhood,

Table 17.1
Organizations Used to Meet Human Needs of a Neighborhood

	*Service Providers***					
	Based Within Neighborhood			*Based Outside Neighborhood*		
*Human Needs and Services**	*Private Nonprofit*	*Private Profit*	*Govern-ment*	*Private Nonprofit*	*Private Profit*	*Govern-ment*
Employment						
Services						
Jobs						
Commercial						
Sale of Goods						
Services						
Income Support Programs						
Credit						
Housing						
Utilities						
Food						
Clothing						
Safety						
Transportation						
Education						
Recreation, Arts						
Health						
Social Services						
Governance						
Civic Participation						
Communal Events						
Religion						

* Detailed services and activities should be added to under each human need.

** Individuals, families, and informal groups will also be responding to human needs.

and many more go outside for employment. Some neighborhood jobs are filled by outsiders. In an affluent, suburban neighborhood, most of the residents who work go elsewhere for their jobs, and not many will be unemployed and seeking work; there are some jobs within the neighborhood for domestic service, lawn and house maintenance, and public services, mostly filled by outsiders. In contrast, poor, inner-city areas usually have a greater variety of jobs within or near the neighborhood but also much higher unemployment. The affluent neighborhood doesn't need an employment program, but the inner-city area requires a combination of counseling, training, placement in private jobs, and public service employment. It should have a neighborhood employment center staffed with outreach workers, counselors, trainers, and job placement specialists. Residents may need to be referred to specialized training institutions outside the neighborhood. Placement personnel will connect the unemployed with jobs and on-the-job training opportunities with private employers both within and outside the neighborhood, and also with jobs with public agencies and nonprofit organizations. In between these two types of neighborhoods are some with different sets of employment needs, such as a blue collar neighborhood where some persons may need retraining to cope with technological change; also, if both parents want to work, day-care services are needed for their children. Another neighborhood may have a number of retired persons who want to supplement their income from social security and a small pension by part-time employment. For them, an appropriate response would be a program to find part-time jobs with private employers and offer public service employment opportunities.

We can make the same analysis for responses to other human needs. When put together, the combined analysis will tell us how effectively each neighborhood and the broader systems with which it is connected are meeting the total needs of the residents. This then forms the basis for remedial action, institutional development, and system reform. There will be a different set of strategies and a different combination of services and programs for each neighborhood, although every neighborhood in some way or other will need a complete array. Previous chapters dealing with self-help activities, municipal and social services, public education, physical preservation and revitalization, economic development, and governance have presented many of the program possibilities from which to choose.

When neighborhood needs are viewed in this manner, the concept of triage — letting some badly deteriorated neighborhoods die — is discarded because even if some housing is beyond saving, the people aren't. However, the poorest neighborhoods will require a different constellation of programs and services than those needed in other kinds of neighborhoods. Thus, completeness has a different meaning for each neighborhood.

FUNCTIONALLY INTEGRATED

But it isn't enough to consider the needs and responses separately. We should also look at the functional connections. For example, being employable requires education and good health. Health depends upon proper nutrition, and to buy food, people need money earned in jobs or obtained through income support programs (or food stamps as equivalent). Personal safety is achieved through fire and police services and also by residents organized as block watches and patrols. Rent and mortgages must be paid from employment earnings, returns from investments, welfare, or some other form of subsidy. In housing projects families may require a full roster of supportive social services. Children in school need good teachers, adequate nutrition, and effective help at home. The latter can be strengthened through neighborhood support systems.

On and on these connections can be enumerated. They occur at two geographic scales: linkages within the neighborhood between various actors and activities, and network and system connections covering a wider territory. We can think of this pair as horizontal and vertical integration.

Horizontal integration occurs within the neighborhood territory in relationships among the various functions: school-home, jobs-education, health-food, housing-social services. Some of this happens naturally, some by necessity, but often the needed connections are not made because of individuality and bureaucratic insularity. Methods to overcome this shortcoming include organizational consolidation, colocation, unified planning, and program coordination.

Conceivably all major neighborhood operations could be consolidated into a single, all-encompassing organization — a neighborhood government or a multipurpose neighborhood corporation. However, because of its size it would have to be divided into operating divisions, just as city government and other large organizations are. Although there would be a line of command with some internal sanctions, problems of coordination among operating units would remain, just as there are within government, business corporations, and private institutions.

Colocation offers some benefits, as we saw in Chapter 12 when we looked at examples of little city halls and multiservice centers. Community schools, mentioned in Chapter 13, can achieve this, too. Having personnel close to one another facilitates communication, which is the first step of functional integration, but it doesn't guarantee it. There still has to be mechanisms to orchestrate cooperative action.

Unified planning is one such mechanism. This can be accomplished by a consolidated organization or by a collection of separate entities coming together to produce a unified plan of activities and capital projects for the neighborhood. There was considerable experience with this approach under the Model Cities Program with success in some locales but also meager results elsewhere. Breakdown usually occurred at the implementation stage, particularly where program operators with independent funding sources could go their own way. Control of fund allocation was an important tool for putting plans into action, and a strong motivator to bring diverse operators into the planning process. Thus, the budget-making process offers one of the best opportunities for unified neighborhood planning, at least to the extent that a number of agencies draw their funds from the same source, such as the city budget, a federal grant, or the United Way.

Program coordination during implementation can rely upon a variety of mechanisms. In earlier chapters, we touched on some of them: coterminous service district boundaries, neighborhood service cabinets, a neighborhood manager in charge of a little city hall, a director at a multiservice center, a community school coordinator, joint intake and case conferences at the multiservice center, a consolidated information and complaint process at the little city hall, network linkages between professional and lay helpers, budget review as a tool to set the conditions for coordination. These mechanisms can be further strengthened by using them to achieve cooperative working relationships among citizens and service delivery personnel. Many years of experience have shown that coordination can be best accomplished at the delivery level, but funding conditions set by legislative acts and the rules and procedures established at agency headquarters make it easier, or harder, to achieve neighborhood coordination.

Vertical integration occurs within functional systems where neighborhood activities are linked with broader operations. This fulfill the system concept which we first considered in Chapter 7 (a group of activities which regularly interact so as to form a unified whole). There is scarcely a neighborhood activity which doesn't have a system relationship. More significantly, many human needs of neighborhood residents can be fulfilled only through these vertical connections. Examples are: neighborhood employment counseling, training in a district vocational school, and placement in jobs outside the neighborhood; referral from the neighborhood health center to a hospital for specialized care; housing rehabilitation supported by mortgages obtained from citywide lending institutions and subsidized by federal grants; meals at senior centers using food obtained from government surpluses and at a central food depository stocking donations from private companies. Such interdependency doesn't diminish the importance of neighborhoods but does place neighborhood action in a true and proper context.

In short, we can investigate both the horizontal and vertical dimensions of functional integration occurring within and beyond particular neighborhoods and then implement measures to remedy shortcomings.

Mending Tears, Healing Wounds

Mending tears in the social fabric and healing psychic wounds of individuals, families, and social groups also enter into achieving neighborhood wholeness. This goes beyond service delivery, though some services contribute to the process.

HELPING NETWORKS

This process can build upon innate neighborliness, personal networks, and natural helpers but may require careful planning and guided application. This occurred, for instance, in the community mental health program in Baltimore and Milwaukee neighborhoods (considered in Chapter 13) where the lay helping network was linked with the professional helping network (look again at Figure 13.1, p. 195). The same concept can be applied to needs of the full range of ages from children to the elderly and to a variety of problems of personal adjustment and social relationships. The key is reciprocal recognition by both lay and professional helpers that the other has something to offer and that they gain by working together. The small scale of the functional neighborhood makes such linkages workable.

DISPUTE RESOLUTION

The neighborhood can also be the seat of conflict resolution. This is especially appropriate because a considerable portion of violence in the United States occurs in a neighborhood setting: child abuse and disputes between family members; youth gangs fighting one another; arguments among neighbors over noise, pets, and property damage (for the neighbor relationship isn't always harmonious); tenants in conflict with landlords and consumers with merchants. Rather than requiring police intervention and ending up in court, many of these situations can be settled through neighborhood justice centers, mediation, and other nonjudicial dispute resolution mechanisms. During the past ten years a number of experiments of this kind have been carried out in a number of cities (McGillis and McGillis, 1977; Cook, et al., 1980; Klein, 1980; Tomasic and Feeley, 1982). Most neighborhoods could benefit from application of these varied techniques of reconciliation.

DEALING WITH INTERGROUP CONFLICT

Intergroup conflict between racial and ethnic groups is another area where neighborhood action is appropriate. Indeed, one source of such conflict is the desire of people to feel secure at home and to protect their property investment. This makes many residents turf-conscious and causes them to feel threatened when somebody "not like us" moves into the neighborhood. We should admit this aspect of neighborhood identity and seek to counter its harmful manifestations.

Earlier I mentioned the desirability of seeking to identify residents' shared interests within the immediate personal neighborhood, their common humanity as parents, children, and citizens, their similar fears and hopes. These similarities are revealed when people communicate with one another and work together to solve common problems. Neighborhood leaders and associations can contribute to better understanding among groups by emphasizing what unites them and thus provide a more peaceful setting in which they can deal with matters which seem to divide.

Philosopher Lawrence Haworth in *The Good City* stressed two key attributes: opportunity and community. He noted (1963:86):

> In a good city, opportunity appears within the context of community. . . .In any genuine community there are shared values: the members are united through the fact that they fix on some object as preeminently valuable. And there is a joint effort, involving all members of the community, by which they give overt expression to their mutual regard for that object.

The value they share may be a goal they consider worth achieving, a memory of a mutual heritage, a common religious faith (1963:20). It can be a social cause, or the neighborhood as a good place to live and improve.

Where two communities of identity — racial, ethnic, or religious — are drawn into conflict, one means of resolution is to focus upon another type of community identity which brings them together. This can be the neighborhood they share. It may not eliminate all opposing values, but working together on matters people agree upon helps to mitigate tensions.

Organizational Alternatives

Organization is needed to make neighborhoods function properly and beneficially to their residents. Here I concentrate on organizations controlled by residents, saving for the next chapter the roles of government, private enterprise, and the nonprofit sector based outside the neighborhood.

FORMS

In previous chapters we have considered a number of organizational forms used in neighborhoods. We can pull them together into six major categories, though recognizing that they might have some overlapping purposes, functions, and strategies:

Neighborhood association: used for two purposes — (1) advocacy or (2) low budget, self-help activities; has individual members, representatives from other organizations, or a combination; ranges from block clubs to multi-issue neighborhood organizations.

Neighborhood congress: an organization of organizations; tends to concentrate upon advocacy and neighborhood organizing but might sponsor a neighborhood corporation for program operations.

Neighborhood advisory committee: usually set up by a governmental agency to deal with issues related to its mission; has an advisory role in program planning and possibly in program implementation and evaluation; selection of members might be by an administrator, nominees of specific organizations and interests, or a combination.

Neighborhood council: unit with official recognition by the city or county to deal with policy issues in a number of program areas; mainly advisory authority, rather than final decision making; members usually elected by residents; in some places existing associations have gained recognition as neighborhood councils.

Neighborhood corporation: funded and staffed to operate specific services or undertake developmental activities; incorporated, usually as a nonprofit organization but might have a profit-making subsidiary (or vice versa).

Neighborhood government: has legal power and authority equivalent to that of a municipality to make policy and run specific programs; has access to financial resources necessary to carry out its responsibilities; democratic selection of governing officials.

Although I've mentioned advocacy only for the first two, all six forms are likely to engage in advocacy as a means of promoting their central missions.

COMBINATIONS

In reviewing the experience of 44 neighborhood organizations, the Task Force on Governance of the National Commission on Neighborhoods remarked that "it is clear from the case studies that there is no one structure that joins all neighborhoods and their needs. Rather, there are many approaches that need to be available, and can be effective in handling neighborhoods' capacity for self-determination and governance" (1979, App. I:57). Frequently a single neighborhood relies upon two or more forms. For instance, the Task Force noted these patterns of dual structures:

Spin-off organizations
- Issue organizations to development and service
- Development corporation to issue organization

Subsidiary corporation
- Independent and city-initiated organizations side-by-side
- Two independent organizations

Some neighborhoods might have a separate corporation for each of several major functions, such as economic development, housing, employment training, health, social services. In other neighborhoods several functions might operate as divisions of an umbrella community development corporation. Not the precise form but the completeness of coverage and the competency of operations are the important factors.

KEY FEATURES

In a survey of grassroots organizations with different emphases, including issue-orientation, service delivery, and economic development, Janice Perlman found that effective organizations had the following characteristics (1979:17):

(1) Full-time, paid professional staff.
(2) Well-developed fund-raising capacity.
(3) Sophisticated mode of operation, including:
 (a) Neighborhood street organizing.
 (b) Advanced issue-research capacity.
 (c) Information dissemination and expose techniques.

(d) Negotiation and confrontation skills.
(e) Management capacity in service delivery and economic development areas.
(f) Policy and planning skills.
(g) Lobbying skills.
(h) Experience in monitoring and evaluating government programs.

(4) Issue growth from the neighborhood to the nation.
(5) A support of umbrella groups, technical assistance, action-research projects, organizer-training schools.
(6) Expanding coalition-building with one another, with public interest groups, and with labor.

Some of the same findings came out of the Urban Institute study of 99 neighborhood development organizations, which we reviewed in Chapter 14 (Marshall and Mayer, 1983). Achieving this level of competency requires time, may need training and technical assistance from outside organizations, and for poorer neighborhoods requires outside financial support.

Neighborhood Government

Because numerous neighborhoods in American cities have demonstrated their operational capability with successful private nonprofit organizations, I believe that neighborhoods should now have the choice of going a next step by organizing neighborhood governments with legal authority. This would enable residents to achieve greater self-governance and take major responsibility for managing service delivery and achieving neighborhood wholeness. Therefore, I offer a model presenting the main features of neighborhood government.

A MODEL

Within the central city, neighborhood government would come into being where municipal government already exists. This differs from the developing suburbs where the suburban municipality starts fresh. The result would be a redistribution of political power from city government to neighborhoods, and this would affect both the structure and operation of neighborhood government. Furthermore, the physical presence of neighborhoods within the city requires very close relationships between these jurisdictions of different size. It should be viewed as a symbiotic relationship, that is, a close association mutually advantageous to both parties.

The source of power for neighborhood government in this model is the city charter, which would set forth the process for establishing neighborhood government, the parameters of authority, and the framework of equitable finance. Details would be filled in by ordinance, and by negotiations of neighborhoods with a designated city official in the creation of the new units. There should be some latitude of neighborhood governmental organization, though with standards of democratic procedures and accountability. A representative governing body appears to be the most workable arrangement, but neighborhoods should be allowed to adopt assembly government if they choose. They should have the option of neighborhood equivalents of mayor-council or council-manager forms, or some fresh form of their devising.

The powers exercised by neighborhood government should put into practice an observation of Thomas Jefferson that "it is not by the consolidation, or concentration of powers, but by their distribution that good government, is effected. . . .It is by this partition of cares, descending in gradation from general to particular, that the mass of human affairs can be best managed, for the good and prosperity of all" (1943:1173). Applying this philosophy, neighborhood units could handle services where small scale is economic, efficient, and most sensitive to personal needs, and city and metropolitan agencies would take care of services where larger scale is necessary. Services would operate as part of functional systems so that neighborhood and city services would be carefully interrelated.

Thus, neighborhood government could handle police patrol and local traffic enforcement while the city would take care of communications, crime laboratory, major criminal investigation, and confinement of felons. The neighborhood could have its own playgrounds while the city recreation department would manage district facilities, such as ice skating rinks and large parks; a metropolitan agency could operate the large sports arena and zoo. The neighborhood could collect trash, and the city or a metropolitan agency would handle disposal. The neighborhood could run a branch library, and the central library would provide the reference collection, facilitate interlibrary loans, and operate a computerized information service. The neighborhood fire company could provide routine fire protection, and the city department would take care of specialized fire fighting, training, communications, pooled service, and back-up arrangements. Neighborhood health centers could offer outpatient services, but their patients and doctors would have access to city hospitals and a regional medical center.

A neighborhood government could have its own economic development department, could sponsor a nonprofit corporation having this responsibility, or could relate to an independent community development corporation. Each of these arrangements has precedents among city governments. Likewise, neighborhood government could sponsor housing projects or work with a separate, nonprofit housing corporation. It could contract various services to other, special-purpose neighborhood corporations, and could even contract with the city, county, or private companies. This, too, has precedent, particularly among suburban municipalities. The essential feature of neighborhood government is policy control through an elected governing body, with many choices available for organizing service delivery (see Rich, 1977).

However, a neighborhood government should have its own tax base over which it has full spending authority. This could be an earmarked share of property, sales, and income taxes levied within its boundaries. But because of the inequitable distribution of both tax base and need, neighborhoods should participate in a system of shared revenues designed to overcome inequities. Some of this redistributed revenue should be general purpose, allowing neighborhood decision on its use, but some could be targeted to specific programs, such as health, housing and community development, employment and training, food and emergency assistance. (We'll look at this matter from a different perspective in the next chapter.)

Although it would concentrate upon governance and service delivery within its territory, a neighborhood government would have an advocacy role in dealing with city, county, state, and national governments and with corporations and institutions in the private sectors. This is similar to city government, which acts as an advocate for its people and its own powers. (For a fuller presentation on neighborhood government, see Hallman, 1974; for other proposals, see references on p. 72.)

FIT WITH AMERICAN SYSTEM

Neighborhood government could fit comfortably into the American governmental system, which is federal in character. That is to say, in the United States we have governmental units functioning in areas of different geographic scale with each unit serving as "different agents and trustees of the people, constituted with different powers, and designated for different purposes" (Madison, n.d.:304-305). In short, one people served by several tiers of government.

Rufus Davis (1967:15) has emphasized three principal features of power and authority which each member of a federal union must possess: (1) independence within its ambit of power, (2) financial means to be master, and (3) political ability to be master. But as Richard H. Leach has pointed out (1970:2), federalism in operation "requires a willingness to cooperate across governmental lines and to exercise restraint and forebearance in the interest of the entire nation."

Neighborhood government would be a unit in a system of local federalism, which also has city and county governments and some type of areawide mechanism (Hallman, 1977b). It would need stable financing and sufficient authority to manage the tasks assigned to it. Elected neighborhood officials, as advocates of neighborhood interests, would function politically in serving as advocates of neighborhood interests in dealing with broader domains. However, they should act in spirit of cooperation and willingness to negotiate with city, county, metropolitan, state, and national agencies to work out adjustments of their respective immediate interests for the benefit of the total community. In this manner neighborhood government would be a natural part of American federalism.

Summation

Neighborhoods evolve, their populations make-up may shift over the years, and the problems they face and the opportunities they possess may also change. The challenge is to use their own resources (human talent as well as economic) and those drawn from the outside to heal wounds, to provide a complete array of needed social, political, economic, and service activities, and to achieve functional integration, both horizontal (within the neighborhood) and vertical (involving broader systems). In short, to strive for neighborhood wholeness.

Residents can choose from a variety of organizational forms: association, congress of organization, advisory committee, officially recognized council or board, a corporation, or their own neighborhood government. Often combinations are useful. The most complete form would be neighborhood government, but even then residents would expect other kinds of organization to function within the neighborhood, and, of course, a variety of informal groups and personal networks.

As residents are pluralistic in their interests and values, so also the clusters of organizations can be diverse. As residents can achieve a heightened sense of community by working together on matters of common concern, so also their varied organizations should strive for unity of purpose on matters of mutual interest, directed toward achieving and maintaining a good neighborhood.

Exercises

(1) Describe the principal attributes of a good neighborhood.

(2) Take one of the neighborhoods you have previously studied and indicate how it does and doesn't measure up to these attributes.

(3) Use Table 17.1 and describe how this neighborhood is organized to meet the basic human needs. Take into consideration organizations in all three sectors, both neighborhood-based and coming from outside the neighborhood.

(4) Describe the methods used to achieve functional integration, both horizontal and vertical.

(5) What new or different organizations and mechanisms would you recommend?

(6) What is being done within this neighborhood to mend tears and heal wounds? What more should be done?

(7) Do you believe that there should be neighborhood government in large cities? Give your reasons for or against.

CHAPTER 18

PARTNERS AND SUPPORTERS

All of us living in urban areas are neighborhood residents, whether or not we have a strong neighborhood identity or are active there. At the same time, many of us have an impact on other people's neighborhoods because of actions we take as elected officials, public employees, corporate officers and employees, board members and employees of private nonprofit organizations, members of associations of many types, and advocates of particular interests. In these capacities, we can be partners with and supporters of neighborhoods, or competitors and adversaries.

Forms of Support

Support for neighborhoods can occur in a number of ways.

DIRECT SERVICES

A wide variety of agencies, organizations, institutions, and private companies provide direct services to neighborhoods, such as the lists of municipal and human services previously presented in Table 12.1 (p. 171) and Table 13.1 (p. 185). Although in a few cities some of these services are contracted to neighborhood organizations, by and large these days most services delivered within city neighborhoods are under the control of agencies of broader domain. To show proper support for neighborhood vitality, these agencies should be expected to perform their tasks effectively, treat residents respectfully, appreciate and respond to differences among neighborhoods, establish neighborhood offices where appropriate, bring residents into advisory roles, and offer jobs to residents when possible.

These expectations aren't always met. Improvements are easier to advocate than accomplish because bureaucracies have tendencies toward standardization and isolation from service recipients. Many professional experts are antagonistic toward, or at best skeptical of, lay participation. As counteraction, top administrators have a responsibility to work on staff attitudes, especially class and racial biases, and bureaucratic practices to make them more in tune with neighborhood needs. Once initiated, actual experience with neighborhood-oriented services is the best teacher, as proven by numerous cases during the past 20 years.

Neighborhood organizations, as interest groups, continuously press for new and better services. Agencies tend to resist, particularly in an era of tight budgets. Although bargaining and balancing varied interests will always be major factors in decisions on resource allocation, agencies have a responsibility to look at relative neighborhood need and to design and assign services equitably. They should also be alert for opportunities for joint endeavors and mobilization of volunteers in ways which stretch available resources.

Private enterprises should likewise pay attention to neighborhoods and their needs in locating stores, hiring personnel, making credit available, and various other economic activities. A narrow focus on the bottom line of profits or losses ignores the civic responsibility of businesses for the welfare of the broader community. Leaving sections of the city devoid of food stores or mortgage funds harms not only those neighborhoods but also the whole community fabric. Businesses, like governmental agencies and nonprofit organizations, should find ways to establish communications with residents and work cooperatively. They should form joint ventures with neighborhood organizations (more on such partnerships later in this chapter).

REFRAIN FROM HARM

Beyond their positive contributions to neighborhoods, agencies, institutions, and private companies should take care that they don't harm neighborhoods by their actions, even unintentionally. In the past, they have, and some are still doing so.

The extension of the interstate highway system into cities has probably done more harm to neighborhoods than any other program of the past 25 years. Urban renewal projects relying on wholesale clearance and excessively large and dense public housing projects have had detrimental effects. Redlining and other disinvestment practices of lending institutions and insurance companies have impaired neighborhood vitality. Decisions by corporate conglomerates and other private companies to shut down manufacturing plants and relocate operations elsewhere have hurt neighborhoods where many of their employees live. Decisions by school boards to close neighborhood schools and by health departments to close hospitals also can have a deleterious effect.

Some neighborhood advocates have proposed that public agencies and private agencies receiving public support should be required to file neighborhood impact statements before making final decisions on new projects and program activities. This would be analogous to environmental impact statements. Personally I'm skeptical of this approach as a universally applied regulatory practice because it could be unduly burdensome and could be made fairly meaningless by glib statement writers. However, carefully focused use, especially for projects which would result in the demolition of housing units, could be beneficial to neighborhood interests.

Even more important is sensitizing planners, administrators, and members of policy boards to the importance of neighborhoods and enhancing their awareness of how their decisions can have either a positive or negative impact.

LEGAL AUTHORITY

Neighborhood organizations require legal authority to operate. If they are nonprofit corporations, their authority stems from the general corporation laws of the state. In most states incorporation in this manner is fairly easy, though legal counsel is usually required. Because many neighborhood corporations wish to qualify as a tax exempt organization under the federal internal revenue code, their articles of incorporation and bylaws must be consistent with the appropriate section of that code. Sometimes they run into obstacles with state government or at the Internal Revenue Service. Officials there should aid rather than resist neighborhood incorporation and tax exemption.

To gain governmental authority is more difficult. The mildest form is as a neighborhood council with advisory power. As previously noted, a number of cities have done this through city charters, ordinances and resolution, and executive orders. But elsewhere proposals have died aborning within charter commission, city councils, and at charter referendums. In states without municipal home rule, the state legislature must act, possibly followed by local approval. The difficulty of such dual action was demonstrated in Indiana where in 1972 the state legislature enacted a statute authorizing the consolidated government of Indianapolis/Marion County (unigov) to set up community councils (minigov), but diluted the prospective powers with a 1973 amendment. However, even then the city-county council stymied implementation by refusing to adopt a boundary plan.

To go the next step and create neighborhood government would require firmer authorization, most likely a charter amendment in home rule states and a legislative act elsewhere. It's quite possible that even in home rule states the legislature would have to enter the picture because of general laws on the powers of municipalities, taxation, and revenue sharing.

Where federal funds are involved, basic acts of Congress and agency regulations should be written to permit grants to or contracts with neighborhood corporations and other nonprofit corporations. Many of them do now, but others have restrictions on this kind of delegation.

REMOVAL OF OBSTACLES

Besides the matter of legal authority, other legal and administrative obstacles hinder neighborhood operations. Some of them relate to the delegation or contracting process, such as whether procurement regulations permit noncompetitive, sole-source contracting with a neighborhood corporation and whether this would be considered an unlawful restraint of trade. Some collective bargaining agreements between unions and local government have clauses prohibiting, or at least impeding, contracting out services now performed by public employees, or they provide for no layoffs. Another matter is the legal liability for harm to persons and property by employees of neighborhood corporations performing a delegated function. In the case of neighborhood health centers, the possibility of malpractice suits must be considered. All these issues require resolution.

Some program operations of neighborhood corporations encounter other kinds of obstacles. For instance, a housing corporation may want to acquire ownership of an abandoned building and rehabilitate it, but runs into complications of tax foreclosure and mortgage foreclosure processes. The city may apply building code standards written for new construction which are too expensive and structurally difficult to achieve in older buildings. If federal funds are involved, the wage standards under the Davis-Bacon Act may raise labor costs beyond what the project can afford and force rents above the low to moderate level (National Commission on Neighborhoods, 1979:174-178, 202-204).

Neighborhood programs involving concurrent operation of several programs (such as occurs at multiservice centers) may encounter overlapping and sometimes conflicting requirements of federal, state, and municipal regulations (such as eligibility criteria, operating procedures, and reporting requirements). Incongruous service district boundaries may be another problem.

Agencies and legislative bodies can support neighborhood action by removing these obstacles.

FINANCIAL RESOURCES

For neighborhood corporations and prospective neighborhood governments to operate effectively, they must have adequate financial resources. Because of unequal distribution of needs and revenue sources, many neighborhoods, particularly poor, inner-city ones, must draw in financial support from outside. This requires methods for redistribution of resources based upon principles of equity.

Equity. By equity, I mean basic fairness. Three sets of outcome can serve as the test (Gans, 1974:63). The first is equality of treatment, especially in matters of citizenship, such as the right to vote and hold office, and equal justice under law. The second is equality of opportunity, which might mean special attention to persons blocked from fair access to opportunity, such as education, employment, and housing. The third is equality of results, which determines whether equal opportunity is really put into practice. Applying these principles might require an unequal input of resources, such as special funds for compensatory education and extra municipal resources in order to achieve similar results in all neighborhoods.

The equity issue comes to the fore in neighborhood government and other forms of neighborhood operations, particularly when considering sources of revenue. Within every city and metropolitan area, the tax base is unequally distributed, whether measured by property values, income, or retail sales, the three principal sources of local revenue. The problem is compounded by inequalities among metropolitan areas, states and regions. It would be inequitable to have every neighborhood rely only upon the tax base present in its territory, for poor neighborhoods wouldn't be able to afford the quality of services richer neighborhoods could provide. The remedy is to supplement neighborhood-based revenues with funds redistributed by broader domains, such as city, county, state, and national, and also private sources. A number of mechanisms are available.

Public grants, loans, contracts. There are currently a variety of federal and state grant funds going into neighborhoods, usually controlled by or filtered through local government. Thus occurs in federal support for local programs of community development, employment and training, compensatory education, health, mental health, social services, and many more. Federal funds are allocated to states and cities on the basis of population and various criteria of need, often with some preference for greatest need and lower-income beneficiaries. State and local recipients in turn are expected to concentrate on local areas of greatest need, though this is often diluted by loose and ambiguous criteria. Nevertheless, from the 1960's until 1981 the overall pattern of federal grants tended to benefit lower income neighborhoods. Then program consolidation into state block grants, proposed by President Reagan and adopted by Congress, reduced the funding level substantially and seriously weakened targeting requirements to the detriment of inner city neighborhoods.

In addition to grants, public programs provide loans for neighborhood projects, such as those we considered in Chapter 14, and some local governments have entered into contracts with neighborhood organizations, as we reviewed in Chapter 12.

The level of federal, state, and local grants, loans, and contracts should be increased, particularly for poorer neighborhoods. Targeting should be more precise, and more grants should go directly to neighborhood organizations without having to filter through state and local government bureaucracies.

Shared revenues. Another method to get more financial resources to neighborhoods would be to bring them into the system of state-local revenue sharing, a practice found in at least one form in all states except Delaware. According to the Advisory Council on Intergovernmental Relations (ACIR), this is done for four reasons (1980:2):

- As a way of compensating localities for property exempted from local taxation or removed from tax rolls.
- To harness the superior revenue raising ability of state tax systems to the local need for more diversified, administerable, and economically responsive revenue sources.
- To provide local property tax relief.
- To respond to the differing need among local governments for revenue and their differing ability to raise it.

The same reasons can be translated to state and local revenues shared with neighborhoods.

In 1977 the states shares $6.6 billion of revenue with local government, divided as follows: (a) 13.0 percent on the basis or origin of revenue, (b) 26.7 as property tax reimbursement, (c) 21.4 percent according to population, (d) 23.1 percent on the basis of tax capacity, tax effort, or other need factors, and (e) the remaining 15.7 percent according to other methods (ACIR, 1980:7). The fourth method, providing the basis for almost one-fourth of state shared revenue, has an equalizing effect. ACIR has made the following case for equalization (1980:9):

> It is unfair for the quality of local public services to be a function of the tax base of the community because this means that residents of tax base rich communities potentially will have high service levels at low tax rates, while those of tax base poor communities will have low service levels or high tax rates. The comparatively poor status of low tax base communities may mean difficulty in attracting new business and residents and may even encourage emigration. Both are circumstances that further erode the tax base and relative attractiveness of the disadvantaged and declining locality.

ACIR further pointed out that the per pupil cost of educating the urban poor and the per capita household cost of providing police, fire, and other services in central cities is greater than the costs in other jurisdictions. Using the same arguments, we can make a strong case for neighborhood participation in revenue sharing.

Tax provisions. A variety of tax provisions help make increased financial resources available to neighborhoods. Tax exempt municipal bonds finance neighborhood facilities and development projects. Property tax abatement encourages housing rehabilitation. The federal internal revenue code contains provisions which offer incentives for high tax bracket persons to invest in certain kinds of housing projects; the depreciation schedule and operating losses result in lower income taxes for the investors. Seven states have adopted laws providing for income tax credits for investments in neighborhood development projects. A growing number of sophisticated neighborhood corporations are taking advantage of the benefits they can gain from these varied tax provisions.

Private sources. Neighborhoods also receive contributions from private sources. Five methods of private giving are used: cash, in-kind donations (property and services), bargain sales, socially responsible investments, and socially responsible contracting (McDonough, Bond & Associates, 1983:11). Foundations and corporations make cash grants, and corporations and individuals donate property and services. Private land owners have entered into bargain sales agreements with community land trusts, permitting them to acquire the land at a price below market value and the owners to gain an income tax deduction. Insurance companies, other corporations, foundations, and churches have used investment funds to facilitate certain kinds of projects, such as for lower-income housing; they receive a return for their investment, though perhaps less than what they might gain from other sources. Pension funds are a vast, untapped potential for neighborhood investments. Developers and other private operators sometimes contract with neighborhood organizations for particular services in order to help create neighborhood jobs.

Summation. Thus, a considerable number of methods are available to provide financial resources to neighborhoods. What is needed is greater initiative on the part of legislators and other policymakers who control sources and a stronger commitment to equity. Neighborhood organizations must maintain persistent pressure to gain the funding they require. For this purpose they need to band together in city, state, and national coalitions (see below).

TECHNICAL ASSISTANCE

Numerous observers of neighborhood organizations have noted that beyond their need for funds they also require technical assistance and training so that they can develop the know-how to plan, implement, and evaluate neighborhood-based activities. Neighborhood organizations themselves recognize this need. For instance, a survey of 76 advanced neighborhood development organizations revealed that their top six technical assistance needs were as follows (Cohen and Kohler, 1983:149):

- Fundraising techniques, funding proposals/grantsmanship, identifying funding services.
- Financial packaging, tax syndications, joint venture housing development.
- Financial management and planning, business accounting.
- Organizational development, board training, management, personnel practices.
- Economic and commercial development, small business development.
- Market analysis, marketing concepts and techniques.

Neighborhood organizations with less experience would need more assistance on basic organizing. Those with different program emphases require technical knowledge of their program fields.

There are a sizable number of technical assistance providers and training institutes functioning at national, regional, state, and local levels (some of them mentioned in Chapter 9). Many are nonprofit organizations with this as their primary function. Universities have centers providing such services. Private entrepreneurs and some fairly sizable corporate enterprises are so engaged. Many state governments have bureaus providing localized technical assistance, and some city and county agencies offer assistance to neighborhood and other nonprofit corporations. In some locales neighborhood alliances have formed their own technical assistance operation.

Federal agencies and foundations have funded some of these providers to offer technical assistance to neighborhood organizations. The providers use their own staff and/or a corps of part-time consultants, some of whom are drawn from more experienced local organizations, thus providing "peer-to-peer" assistance. Locally sometimes corporations loan executives on a part-time basis or for an extended period. Some federal, state, and foundation grants permit neighborhood organizations to include technical assistance funds in their budgets so that they themselves can decide who they want to retain for this service.

If done properly, expenditure for technical assistance and training the staff, boards, and other volunteers of neighborhood organizations is an excellent investment.

Partnerships

Because neighborhoods don't stand alone and don't always have the financial resources and talent they need to solve problems and provide the quality of life the residents want, they need to enter into partnership with outsiders. This is a common phenomenon and takes a variety of forms.

NEIGHBORHOOD/PRIVATE SECTOR

Table 18.1 presents a typology of partnerships which neighborhoods have formed with the private sector, particularly in relationship to developmental activities, according to scholars from Neighborhood Policy Research in Cambridge, Massachusetts.

The first main category is *investment partnerships* "in which both the community group and the private partner have made substantial commitments of resources to a project and both partners have a substantial degree of control over project direction and outcome" (Bratt et al., 1983:24-30). One type is a joint venture, carried out through a new legal entity. An example is the River East Economic Revitalization Corporation in Toledo which joined with a local supermarket chain to develop a mini-shopping center. Another type occurs when a private

Table 18.1
Types of Neighborhood/Private Sector Partnerships

A. Investment Partnerships
   1. Joint ventures
   2. Direct private investments
   3. Syndication
   4. Subsidiary formation

B. Supportive Partnerships
   1. Direct provision of financial resources
      a. Loans and grants to community groups
      b. Loans and grants to neighborhood residents
   2. Cooperation
   3. Technical and organizational assistance
   4. Volunteer participation
   5. Loan guarantees

C. Contracts
   1. Contracts for services to the community
   2. Contracts supporting community institutions and efforts

SOURCE: Bratt et al., 1983: 23.

investor makes an equity investment in a community-based venture, such as has occurred with the Illinois Neighborhood Development Corporation, which we discussed in Chapter 15. A third type is syndication in which wealthy individuals combine with sponsors of developments for the tax benefits as, for example, has been done by the South East Bronx Community Organization in New York and by the Brightwood Development Corporation in Springfield, Massachusetts. The fourth approach is subsidiary formation wherein a nonprofit neighborhood corporation forms a for-profit subsidiary which serves as an intermediary in the neighborhood's relationship with the private sector. An example is the De Sales Restoration Corporation in St. Louis, which is a vehicle for receiving investment support from the Tower Grove Bank and Trust Company, located in that community.

A second main category consists of *supportive partnerships,* functioning cooperatively but without the legal relationships often found in investment partnerships (Bratt et al., 1983:30-42). One approach is the direct provision of financial resources through loans and grants to neighborhood organizations or to individual residents. The former has occurred in Minneapolis where the Dayton-Hudson Foundation made a $1 million grant to the Whittier Alliance for planning and implementing a neighborhood revitalization program (Hanson and McNamara, 1981). The Tasty Baking Company in Philadelphia has supported an organization called the Allegheny West Foundation, serving the community around its plant, and has received tax benefits under the Pennsylvania Neighborhood Preservation Act. Neighborhood housing services (which we studied in Chapter 12) serve as channels for loans from lending institutions to homeowners. A second type of supportive partnership consists of cooperative efforts to achieve common, socially oriented goals. Examples include a youth shelter in Greenwich, Connecticut, wood and fish cooperatives in Portland, Maine, commercial revitalization in the Westport area of Kansas City, Missouri, and the sale of abandoned homes on the North Side of Pittsburgh, Pennsylvania. Technical and organizational assistance is a third approach. For example, the Tucson (Arizona) Barrio Association has benefited from executives loaned from local corporations. The Bedford-Stuyvesant Restoration Corporation receives professional counsel through a parallel organization which draws in corporate executives. Fourthly, volunteers contribute to the survival of many neighborhood organizations, such as those helping Operation Brotherhood in Chicago with food programs. In Memphis volunteers recruited through the Catholic diocese and trade unions have helped an organization called CoDe North carry out housing rehabilitation. Loan guarantees are a fifth method of supportive partnerships, such as the Ford Foundation made to help the Bedford-Stuyvesant Restoration Corporation convert an old dairy into a shopping center.

The third category of neighborhood/private sector partnerships comes in the form of *contracts* (Bratt et al., 1983:42-44). One type is contracts for services to the community, as illustrated New York's Management in Partnership Program whereby community partners are paired with private firms which oversee neighborhood management of previously abandoned properties. A second type is contracts in support of community institutions and efforts, such those as the Humboldt Construction Company, a for-profit subsidiary of a community development corporation in Chicago, enters into with private individuals and social service agencies.

These examples from the Neighborhood Policy Research scholars have emphasized development activities and involvement of the private enterprise sector. In addition, the private nonprofit sector based outside neighborhoods also works closely with neighborhood organizations on a variety of concerns related to basic human needs. The United Way and its affiliated agencies serve neighborhoods and work with neighborhood organizations, which in some cities have been able to gain United Way funding. Private foundations give grants to neighborhood corporations. So do church denominational boards and affiliates, such as the

Campaign for Human Development, funded by the Roman Catholic Church. Local churches support many neighborhood activities, and so do some service clubs. (See Meyer, 1982 for examples of private sector initiatives to solve community problems, sometimes in partnership with neighborhood organizations, and see Weber and Lund, 1981 on the role of corporations in the community.)

Nationally a number of intermediary organizations are channeling foundation and private corporations to neighborhood organizations. One of them is the Local Initiatives Support Foundation (LISC), a spin-off of the Ford Foundation and headed by Mitchell Sviridoff, which works directly with neighborhood corporations and through local committees of private sector interests in geographic areas of concentration. Another is the Enterprise Foundation, organized by developer James Rouse. A third is the Inner City Ventures Fund, a unit of the National Trust for Historic Preservation. Although not directly involved in funding, the Center for Corporate Public Involvement advises life and health insurance companies on ways they can contribute to community solutions through social investments and employee voluntarism.

PUBLIC/NEIGHBORHOOD AND TRILATERAL

There are numerous examples of public/neighborhood partnership, as discussed in previous chapters: community schools, joint neighborhood service centers, police teams working with block watches, coproduction of public services, neighborhood planning, community development projects, and many more. The various mechanisms through which public funds flow to neighborhood organizations provide another expression of partnership. Also, citizen participation processes.

Another form of public/private sector partnership functions in many communities where local government and corporate interests sponsor joint activities and major development projects (CED, 1982; SRI, 1982; R. Berger, 1982; President's Task Force on Private Sector Initiatives, 1982). Most of these are often citywide in scope and have no place for neighborhood organizations, though in some places they do.

A conscious effort to bring neighborhoods into such endeavors occurred during 1980 to 1983 under a trilateral partnership project carried out in a dozen cities under the joint sponsorship of the U.S. Conference of Mayors, the Charles Stewart Mott Foundation, and the U.S. Department of Housing and Urban Development. Emphasis was upon two types of projects: (a) real estate or business development projects requiring some public support or leveraging to be feasible in a market context, and (b) neighborhood or small business wealth-producing activities requiring some front-end support to attain levels of self-sufficiency. Experience in the demonstration indicated that a trilateral partnership needs an on-going organizational structure to provide continuity and a means of communication. Representatives of each sector should have equal standing and weight. In economic development projects neighborhood organizations should have an equity position. Most of the neighborhood organizations required technical assistance in order to play equal role in the partnership.

The Committee for Economic Development, after studying partnerships involving all three sectors — government, private enterprise, and non-profit organizations (including neighborhood organizations in some instances) — concluded that they must be built upon solid civic foundations with the following features (1982:2):

- A civic culture that fosters a sense of community and encourages citizen participation rooted in practical concern for the community.
- A commonly accepted vision of the community that reorganizes its strengths and weaknesses and involves key groups in the process of identifying what the community can become.
- Building-block organizations that blend the self-interests of members with the broader interests of the community and translate those mutually held goals into effective action.
- A network among key groups that encourages communication and facilitates the mediation of differences.
- Leadership and the ability to nurture "civic entrepreneurs," that is, leaders whose knowledge, imagination, and energy are directed toward enterprises that benefit the community.
- Continuity in policy, including the ability to adapt to changing circumstances, that fosters confidence in sustained enterprises.

As important as partnerships are for neighborhoods, they should be on guard not to be coopted and not have their advocacy mission harmfully dulled. Although neighborhood advocates participating in partnership endeavors may need to temper their rhetoric in interest of good relationships and be willing to compromise in order to achieve gains, even if not perfection, for their neighborhood, yet their advocacy should remain clear and firm. Any meaningful partnership has latitude for honest expression, which can open issues requiring resolution and thereby set the stage for seeking mutual understanding and cooperative solutions.

Alliances of Neighborhoods

As neighborhoods need supporters and partners, so also they need each other. Neighborhood organizations function in a pluralistic world where interests compete for attention, power, and share of resources. Residents themselves are pluralistic in their interests. In this atmosphere, neighborhoods can often get more accomplished by banding together in coalitions and alliances. This has occurred in some cities, a few states, and nationally to some extent, but not as effectively as could and should happen.

LOCAL

In studying coalitions of grassroots groups, of which the neighborhood coalition is one variety, Hans B.C. Spiegel has identified a number of earmarks (1981):

- A grassroots coalition is an organization of local organizations that usually band together at first to achieve a specific limited objective.
- There is a refreshing bottom-up quality about these coalitions.
- The issues with which a coalition deals usually grow out of local or special interest concerns. The issues are "nitty gritty," often parochial, reflecting down-to-earth problems that the constituent groups are experiencing.
- One of the central purposes of a coalition is to be an effective protest and negotiation mechanism.
- However, the repertoire of methods through which coalitions perform their tasks is much more varied than protest and negotiation.

Spiegel indicated that to function effectively the coalition must first be organized, an effort that in some cases has taken years. Accordingly, the services of a professional organizer may be useful. Fundraising must be a key concern, for coalitions often live in a state of uncertainty about the money needed for their operations. Considerable research is required for selection and analysis of issues, and this may require staff assistance. The membership needs to educate itself on many items, and the coalition must make linkages with selected outside organizations. Marilyn Gittell, in a study of community organizations in education, has also stressed the importance of networking with other organizations and sources of technical assistance (1980:115-136).

My observation is that coalitions often deliberately limit their agenda to matters on which they can gain a reasonable consensus and avoid issues which divide the members. However, there are instances where citywide coalitions of neighborhood organizations have taken up such issues as the allocation of federal community development funds, for which many of the member organizations are competing, and used the coalition as a forum to negotiate an agreement of a fair share for all neighborhoods. This then strengthened their position as they contended with other interests trying to tap the same funds.

For instance, in the spring of 1983 the Pittsburgh Neighborhood Alliance worked out a scheme to use special, federal job creation funds for neighborhood commercial rehabilitation and development projects. The mayor agreed to $500,000 for this purpose and promised a like amount from the city's next capital improvement budget. In this manner the Alliance produced a new pot of money all neighborhoods can draw upon when they come up with good projects.

Working out this kind of agreement and sticking together over the long haul isn't easy for neighborhoods to achieve, though it is possible. Of all local organizations related to the neighborhood movement, the neighborhood coalition is the most fragile.

STATE AND NATIONAL

Even more difficult is the formation of state and national coalitions of neighborhoods. That's because neighborhood organizations are overwhelmingly locality oriented, even though state and national policies may have serious impact upon neighborhoods. In a few states, such as California, Utah, and Pennsylvania, there have been statewide conferences of neighborhood organizations and other persons interested in neighborhoods, but not any strong, ongoing state coalition of neighborhoods.

The National Neighborhood Coalition, described in Chapter 10, is an alliance of national organizations and technical assistance providers which are concerned about neighborhoods, particularly low and moderate income areas. Some of them have a grassroots membership and most of the others have a grassroots constituency of neighborhood organizations they assist, but the lines of communication tend to flow mostly outward from Washington, D.C., rather than emanate from localities around the nation. But the National Neighborhood Coalition makes a useful contribution by vocalizing neighborhood interests in federal legislation and administrative policies which impact upon neighborhoods and by pushing for more resources, especially for low- and moderate-income neighborhoods, whose needs are greatest.

Local, state, and national coalitions of neighborhoods are all important and should be strengthened.

A Final Word

I started this book by claiming that the neighborhood is a natural phenomenon. In successive chapters I showed how this takes expression as personal arenas of several sizes, as social communities of personal networks and institutions, as a physical place, a political community, and a little economy. In Part II I reviewed one hundred years of history of efforts to shape and improve neighborhoods, and in Part III I described a wide range of neighborhood action in many, varied fields of service and enterprise. Throughout the book, in many different ways, I have insisted that neighborhoods are but one part of urban society and that their vitality depends upon linkages with people, institutions, and processes in the broader community. Most neighborhood residents express these connections in many different ways in their own lives.

Although neighborhoods are natural, the survival and viability of any particular neighborhood doesn't come automatically. It occurs because the residents care. Or as one neighborhood organization tells the residents, "If you want to live in a better neighborhood, join our association and work with us." But no neighborhood can survive alone, not even the newest and most prosperous. Rather all neighborhoods require the care and concern of elected officials, governmental employees, personnel of voluntary agencies, persons holding positions in private enterprises, and citizens who care about other people's neighborhoods as well as their own.

Neighborhoods have long had an important place in American life. They will continue to exist, but the quality of neighborhood life will depend upon what all of us do to preserve and continuously revitalize this precious heritage.

Exercises

(1) Take your neighborhood, or another one you have studied, describe the kinds or outside support it receives.
(2) Describe the actors involved in providing this support.
(3) On the issue of equity, do you consider it proper to redistribute resources for the benefit of lower income neighborhoods? Why or why not?
(4) Find and describe a partnership arrangement between a neighborhood and government or private sector. What has it accomplished? What are its shortcomings?
(5) Look for an alliance between two or more neighborhoods. Describe and evaluate its activities.

(6) What do you as an individual intend to do now and in the future to contribute to achieving and preserving good neighborhoods?

REFERENCES

ABRAHAMSON, J. (1959) *A Neighborhood Finds Itself.* New York: Harper & Row.

ABU-LUGHOD, J. and FOLEY, M.M. (1960) "Consumer strategies" in Foote, Abu-Lughod, Foley, and Winnick, *Housing Choices and Constraints.* New York: McGraw-Hill.

ADDAMS, J. (1910) *Twenty Years at Hull House.* New York: Macmillan.

——— (1930) *The Second Twenty Years at Hull House.* New York: Macmillan.

Advisory Commission on Intergovernmental Relations [ACIR] (1967) *1967 State Legislative Program.* Washington: Government Printing Office.

——— (1972) *The New Grass Roots Government: Decentralization and Citizen Participation in Urban Areas.* Washington: Government Printing Office.

——— (1974) *Governmental Functions and Processes: Local and Areawide.* Washington: Government Printing Office.

——— (1979) *Citizens Participation in the American Federal System.* Washington: Government Printing Office.

———(1980) *The State of State-Local Revenue Sharing.* Washington: Government Printing Office.

AGRANOFF, R. (1977) "Services integration" in Anderson, Frieden, and Murphy, *Managing Human Services.* Washington: International City Management Association.

AHLBRANDT, R. S. , Jr., and BROPHY, P. C. (1975) *Neighborhood Revitalization: Theory and Practice.* Lexington, MA: Lexington Books.

AHLBRANDT, R. S., Jr., and CUNNINGHAM, J. V. (1979) *A New Public Policy for Neighborhood Preservation.* New York: Praeger.

——— (1980) *Pittsburgh Residents Assess Their Neighborhoods.* Pittsburgh: School of Social Work, University of Pittsburgh.

ALDRICH, H. (1975) "Ecological succession in racially changing neighborhoods: a review of the literature." *Urban Affairs Quarterly* 10, no. 3 (March): 327-348.

ALESHIRE, R. A. (1970) "Planning and citizen participation: costs, benefits, and approaches." *Urban Affairs Quarterly* 5, no. 4 (June): 369-393.

ALINSKY, S. D. (1969), *Reveille for Radicals,* (1946). New York: Vintage Books.

——— (1971) *Rules and Radicals.* New York: Random House.

ALTSHULER, A. A. (1970) *Community Control: The Black Demand for Power in Large American Cities.* New York: Pegasus.

ANDERSEN, W. G. , Jr., SPIEGEL, H. B. C., SUESS, A, and WOODS, W. K. (1979) *Profiles of Participation: A Workbook on Citizen Orgnization & Action.* New York: National Municipal League.

ANDERSON, B. E. (1976) *The Opportunities Industrialization Centers: A Decade of Community-based Manpower Services.* Philadelphia: Wharton School, University of Pennsylvania.

ANTUNES, B. GARTZ, C. M. (1975) "Ethnicity and participation: a study of Mexican-Americans, blacks and whites." *American Journal of Sociology* 80 (March): 1192-1211.

APPLEYARD, D. with GERSON, M. S. and LINTELL, M. (1981) *Livable Streets.* Berkley, CA: University of California Press.

ARNAUDO, P. S. A. and PEEL, T. (1974) *Citizen Participation in the Executive Budget Process.* Washington: International City Management Association.

ARNSTEIN, S. R. (1969) "A ladder of citizen participation." *Journal of the American Institute of Planners* 35, no. 4 (July): 216-224.

BAILEY, R. Jr. (1974) *Radicals in Politics: The Alinsky Approach.* Chicago: University of Chicago Press.

BALTZELL, E. D. (1958) *Philadelphia Gentlemen: The Making of a National Upper Class.* New York: Free Press.

BARNES, J. A. (1954) "Class and committee in a Norwegian island parish" in *Human Relations* 7:39-58.

BARTON, A. H. et al. (1977) *Decentralizating City Government: An Evaluation of the New York City District Manager Experiment.* Lexington, MA: Lexington Books.

BATKO, W., CONNOR, P., AND TAYLOR, J. (1975) *The Adams-Morgan Business Sector.* Washington: Institute for Loacl Self-Reliance.

BAUER, C. (1945) "Good neighborhoods." *Annals of the American Academy of Political and Social Science.*

BEAM, K. S. (1935) "The coordinating council movement." *1935 Yearbook of National Probation Association.* New York: The Association.

——— (1936) "Community coordination for prevention of delinquency," in *1936 Yearbook of National Probation Association.* New York: The Association.

BECK, B. M. (1969) "Community control: a distraction, not an answer." *Social Work* 14 (October): 14-20.

BELL, W. and BOAT, M. T. (1957) "Urban neighborhoods and informal social relations." *American Journal of Sociology* 62, no. 4 (June); 391-398.

BENSON, C. S. and LUND, P. (1969) *Neighborhood Distribution of Local Public Services.* Berkley, CA: Institute of Government Studies.

BERG, M. and MEDRICH, E. A. (1980) "Children in four neighborhoods: the physical environment and its effect on plan and play patterns." *Environment and Behavior* 12, no. 3 (September): 320-348.

BERGER, B. (1960) *Working Class Suburb: A Study of Auto Workers in Suburbia.* Berkeley: University of California Press.

BERGER, P. and NEUHAUS, R. (1977) *To Empower People: The Role of Mediating Structures in Public Policy.* Washington: American Enterprise Institute.

BERGER, R. A., [ed.](1982) *Public-Private Partnerships in American Cities: Seven Case Studies.* Lexington, MA: Lexington Books.

BERRY, D. E. and BELL, G. (1978) *Catalysts for Neighborhood Economics.* Kettering, OH: Charles F. Kettering Foundation

BAIGGI, B. (1978) *Working Together: A Manual to Help Groups Work More Effectively.* Amherst, MA: Citizen Involvment Training Program, University of Massachusetts.

BIRCH, D. L. et al. (1979) *The Behavioral Foundations of Neighborhood Change.* Cambridge, MA: Joint Center for Urban Studies.

BIRD, R. N. (1981) *Neighborhood-Based Service Delivery.* Washington: Public Technology, Inc.

BISH, R. L. and OSTROM, V. (1973) *Understanding Urban Government: Metropolitan Reform Reconsidered.* Washington: American Enterprise Institute.

BLECHER, E. M. (1972) *Advocacy Planning for Urban Development — With Analysis of Six Demonstration Programs.* New York: Praeger.

BLOCH, P. B. and SPECHT (1973) *Neighborhood Team Policing.* Washington: National Institute of Law Enforcement and Criminal Justice, U.S. Department of Justice.

BLUMENFELD, H. (1979) *Metropolis . . . and Beyond.* P. Spreiregen (ed.) New York; John Wiley.

BOTT, E. (1957) *Families and Social Networks.* London: Tavistock.

BOURNE, L. S., [ed.] (1982) *Internal Sructure of the City; Readings on Urban Form, Growth and Policy.* New York: Oxford University Press.

BOWEN, D. R. and MASOTTI, L. H. (1968) "Spokesman for the poor: an analysis of Cleveland's poverty board candidates." *Urban Affairs Quarterly* 4, no. 1 (September): 89-110.

BOWSHER, P. (1980) *People Who Care: Making Housing Work for the Poor.* Washington: Prentice Bowsher Associates.

BOYTE, H. C. (1980) *The Backyard Revolution.* Philadelphia: Temple University Press.

BRADBURN, N. and GOCKEL, G. (1971) *Side by Side: Integrated Neighborhoods In America.* Chicago: Quadrangle Books.

Brandeis University, Florence Heller School for Advanced Studies in Social Welfare (1969) *Community Representation in Community Action Programs.* Final Report. Waltham, MA: Florence Heller School.

BRATT, R., BYRD, M., and HOLLISTER, R. M. (1983) *The Private Sector and Neighborhood Preservation.* Washington: Office of Policy Development and Research, U.S. Department of Housing and Urban Development.

BRAZIER, A. M. (1969) *Black Self-Determination: The Story of the Woodlawn Organization.* R. G. and R. F. DeHaan (ed.) Grand Rapids, MI: Willaim B. Eerdmans Publishing Co.

BRODEN,T., SWARTZ, T., ROOS, J. J., and KIRKWOOD, R. B. (1979) *Neighborhood Identification Handbook.* South Bend, IN: Institute for Urban Studies, University of Notre Dame.

BROWN, L. (1977) "Community relations," in B. L. Garmire (ed.) *Local Government Police Management.* Washington: International City Management Association.

BRUDNEY, J. L. and ENGLAND, R. E. (1983) "Toward a definition of the coproduction concept." *Public Administration Review* 43, no. 1 (January/February): 59-65.

BUELL, E. M., Jr. (1982) *School Desegregation and Defended Neighborhoods.* Lexington, MA: Lexington Books.

BURGESS, E. W. (1925) "The growth of the city," in R. E. Park, E. W. Burgess, and R. D. McKensie *The City.* Chicago: University of Chicago Press.

BURKE, E. M. (1966) "Citizen participation in renewal." *Journal of Housing* 23, no. 1 (January): 20.

BUSH-BROWN, L. (1969) *Garden Blocks for Urban America.* New York: Scribner.

CAFTEL, B. J. (1978) *Community Development Credit Unions: A Self-Help Manual.* Berkeley, CA: National Economic Development Law Project.

CAHN, E. S. and PASSETT, B. A. [eds.] (1970) *Citizen Participation: A Casebook in Democracy.* Trenton: Community Action Training Institute.

California Office of Appropriate Technology, Community Assistance Group (1981) *Working Together: Self-Reliance in California's Communities.* Sacramento: Author.

CAPLAN, G. (1974) *Support Systems and Community Mental Health.* New York: Behavioral Publications.

——— and KILLILEA, M. (1976) *Support Systems and Mutual Aid.* New York: Grune & Stratton.

CAPLOW, T. and FORMAN, R. (1950) "Neighborhood interactions in a homogenous community." *American Sociological Review* 15: 357-366.

CARLSON, K. (1978) *New York Self Help Handbook: A Step by Step Guide to Neighborhood Improvement.* New York: Citizens Committee for New York City.

CARMICHAEL, S. and HAMILTON, C. V. (1967) *Black Power: The Politics of Liberation in America.* New York: Vintage Books.

CASSIDY, R. (1980) *Livable Cities: A Grass-Roots Guide to Rebuilding Urban America.* New York: Holt, Rinehart, & Windston.
Center for Governmental Studies (1974) *Charter Language and Ordinances on Neighborhood Decentralization.* Washington: Author.
Center for Neighborhood Development (1980) *Neighbor: Investing in Your Neighborhood and Cooperating with Your Neighbors.* Kansas City, MO: The Center and Park College School for Community Education.
Center for Neighborhood Technology, *The Neighborhood Works (monthly bulletin). Chicago: Author.*
CHECKOWAY, B. (1981) "Politics of public hearings." *Journal of Applied Behavioral Science* 17, no. 4: 566-582.
——— and VAN TIL, J. (1978) "What do we know about citizen participation?" in S. Langton (ed.) *Citizen Participation in America.* Lexington, MA: Lexington Books.
CHURCHILL, H. (1945) *The City is the People.* New York: Harcourt Brace Jovanoritch.
Citizens Planning and Housing Association (1982) *CPHA's Baltimore Neighborhood Self Help Handbook.* Baltimore: Author.
Civic Action Institute [CAI] (1979a) *Childrens' Advocacy.* Washington: Author.
——— (1979b) *Community-Based Waste Recyclying.* Washington: Author.
——— (1979c) *Neighborhood Food Programs.* Washington: Author.
——— (1979d) *Neighborhood Services for the Aging.* Washington: Author.
——— (1980) *Organizing to Handle Neighborhood-Based Technologies.* Washington: Author.
——— *Neighborhood Ideas* (a bulletin issued ten times each year). Washington: Author.
CLARK, K. B. (1965) *Dark Ghetto.* New York: Harper & Row.
——— and HOPKINS, J. (1969) *A Relevant War Against Poverty: A Study of Community Action Programs and Observable Social Change.* New York: Harper & Row.
CLAY, P. L. (1979) *Neighborhood Renewal: Middle Class Resettlement and Incumbent Upgrading in American Neighborhoods.* Lexington, MA: Lexington Books.
——— (1981) *Neighborhood Partnerships in Action.* Washington: Neighborhood Reinvestment Corporation.
CLOWARD, R. A. and OHLIN, L. E. (1960) *Delinquency and Opportunities: A Theory of Delinquent Groups.* New York: Free Press.
COHEN, R. (1978) *Partnerships for Neighborhood Preservation: A Citizens' Handbook.* Harrisburg, PA: Pennsylvania Department of Community Affairs.
——— (1979) "Neighborhood planning and political capacity." *Urban Affairs Quarterly* 14, no. 4 (March): 337-362.
——— and KOHLER, M. (1983) *Neighborhood Development Organization after the Federal Funding Cutbacks.* Washington: Office of Policy Development and Research, U.S. Department of Housing and Urban Development.
COLBURN, F. (1963) *Neighborhoods and Urban Renewal.* New York: National Federation of Settlements and Neighborhood Centers.
COLE, L. A. (1976) "Comment on 'black perspective'" [see A. K. Karnig]. *Urban Affairs Quarterly* 12, 2 (December): 243-250.
COLE, R. L. (1974) *Citizen Participation in the Urban Policy Process.* Lexington, MA: Lexington Books.
——— (1981) "Participation in community service organizations." *Journal of Community Action* 1, no.1 (September/October): 53-60.
COLEMAN, R. P. (1978) *Attitudes Towards Neighborhoods: How Americans Choose to Live.* Cambridge, MA: Joint Center for Urban Studies.

COLLINS, A. and PANCOAST, D. (1976) *Natural Helping Networks: A Strategy for Prevention.* Washington: National Association of Social Workers.

COMER, J. P. (1980) *School Power: Implications of an Intervention Project.* New York: Free Press.

——— (1982) "Parent participation: a key to school improvement." *Citizen Action in Education* 9, no. 2 (November).

Committee for Economic Development [CED] Research and Policy Committee (1970) *Reshaping Government in Metropolitan Areas.* New York: Author.

——— (1982) *Public-Private Patnership: An Opportunity for Urban Communities.* New York: Author.

COOK, Robert F. et al. (1981) *Public Service Employment in Fiscal Year 1980.* Princeton, NJ: Urban and Regional Research Center, Princeton University.

COOK, Royer F. and ROEHL, J. A. (1983) *Neighborhood-Based Crime and Arson Prevention Efforts.* Washington: Office of Policy Development and Research, U.S. Department of Housing and Urban Development.

——— and SHEPPARD, D. J. (1980) *Neighborhood Justice Center Field Test.* (final evaluation report). Washington: National Institute of Justice, U.S. Department of Justice.

COOLEY, C. H. (1909) *Social Organization: A Study of the Larger Mind.* New York: Scribner.

COOPER, T. L. (1980) "Bureaucracy and community organization: the metamorphosis of a relationship." *Administration and Society* 11, no. 4 (February): 411-444.

COX, F. M., ERLICH, J. L., ROTHMAN, J, and TROPMAN, J. E., [eds.] (1974) *Strategies of Community Organization: A Book of Readings.* Itasca, IL: F.E. Peacock Publishers.

CRENSON, M. A. (1978) "Social networks and political processes in urban neighborhoods" in *American Journal of Political Science* 22, no. 3 (August): 578-594.

CUNNINGHAM, J. V. (1965) *The Resurgent Neighborhood.* Notre Dame, IN: Fides Publishers.

——— (1979) *Evaluating Citizen Participation: A Neighborhood Organizer's View.* Washington: Civic Action Institute.

——— and KOTLER, M. (1983) *Building Neighborhood Organizations: A Guidebook.* Notre Dame, IN: University of Notre Dame Press.

DAHIR, J. (1947) *The Neighborhood Unit Plan: Its Spread and Acceptance: A Selected Bibliography with Interpretative Comments.* New York: Russell Sage Foundation.

DAVIDOFF, P. (1965) "Advocacy and pluralism in planning." *Journal of the American Institute of Planners* 31, no. 4 (November): 331-338.

DAVIDSON, J. L. (1979) *Political Partnerships: Neighborhood Residents and Their Council Members.* Beverly Hills, CA: Sage Publications.

DAVIES, C. J., III (1966) *Neighborhood Groups and Urban Renewal.* New York: Columbia University Press.

DAVIES, D. (1978) *Citizen Participation in Education.* Boston: Institute for Responsive Education.

——— [ed.] (1981) *Communities and Their Schools.* New York: McGraw-Hill.

DAVIS, A. F. (1967) *Spearheads for Reform: The Social Settlements and the Progressive Movement, 1890-1914.* New York: Oxford University Press.

DAVIS, R. M. and STEA, D. (1977) *Maps in Minds: Reflections on Cognitive Mapping.* New York: Harper & Row.

DAVIS, R. (1967) "The 'federal principle' reconsidered," in A. Wildavsky, (ed.) *American Federalism.* Boston: Little, Brown.

DEAN, J. T. (1958) "The neighborhood and social relations" in *Forum on Neighborhoods Today and Tomorrow* no. 3 (April). Philadelphia: Philadelphia Housing Association.

DE TOQUEVILLE, A. (n.d.) *Democracy in America* (translated by Henry Reeve) New York: A. S. Barnes & Co.

DE TORRES, J. (1972) *Governmental Services in Major Metropolitan Areas.* New York: The Conference Board.

DILLICK, S. (1953) *Community Organization for Neighborhood Development: Past and Present.* New York: Woman's Press and William Morrow & Co.

District of Columbia Government, Office of Budget and Management Systems (1978) *Improving Productivity in Washington, D.C. Neighborhoods: A Case Study.* Washington: U.S. Department of Housing and Urban Development.

DOOLITTLE, R. J. and MACDONALD, D. (1978) "Communication and a sense of community in a metropolitan neighborhood: a factor analytic examination." *Communications Quarterly* 26, no. 3 (Summer): 2-7.

DOWDEN, C. J. (1980) *Community Associations: A Guide for Public Officials.* Washington: Urban Land Institute.

DOWNS, A. (1973) *Opening Up the Suburbs: An Urban Strategy for America.* New Haven: Yale University Press.

——— (1975) "A basic perspective concerning the community development program" in *Recommendations for Community Development Planning.* Chicago: Real Estate Research Corporation.

——— (1981) *Neighborhoods and Urban Development.* Washington: Brookings Institution.

DRAKE, St. C. and CAYTON, H. (1945) *Black Metropolis.* New York: Harper & Row.

——— (1965) "The social and economic status of the Negro in the United States" in *Daedalus* (Fall): 771-814.

EAMES, E. and GOODE, J. G. (1977) *Anthropology of the City: An Introduction to Urban Anthropology.* Englewood Cliffs, NJ: Prentice-Hall.

EDDY, W., PAPP, S, and GLAD, D. (1970) "Solving problems in living: the citizens viewpoint." *Mental Hygiene* 54: 64-72.

EHRLICH, P. (1979) *Mutual Help for Community Elderly: The Mutual Help Model, vol. 2, Handbook for Developing a Neighborhood Group Program,* Carbondale: Rehabilitation Institute, Southern Illinois University.

EISENBERG, P. (1981) "Citizen monitoring of block grants." *Journal of Community Action* 1, no. 1 (September/October): 4-11.

FAINSTEIN, N. S. and FAINSTEIN, S. (1976) "Future of community control." *American Political Science Review* 70, no. 3 (September): 905-923.

FAINSTEIN, N. S. and MARTIN, M. (1978) "Support for community control among local urban elites." *Urban Affairs Quarterly* 13, no. 4 (June): 443-468.

FALKSON, J. (1974) *An Evaluation of Policy Related Research on Citizen Participation in Municipal Service Systems; Overview and Summary.* Washington: TARP Institute.

FANTINI, M. D., GITTELL, M., and MAGAT, R. L. (1970) *Community Control and the Urban School.* New York: Praeger.

FAUX, G. (1971) "Background paper" in *CDCs: New Hope for the Inner City* (report of the Twentieth Century Fund Task Force on Community Development Corporations). New York: Twentieth Century Fund.

FESTINGER, L., SCHACTER, S., AND BACK, K. (1950) *Social Pressures in Informal Groups.* New York: Harper & Row.

FISCHER, C. S. et al. (1977) *Networks and Places: Social Relations in the Urban Setting.* New York: Free Press.

FISH, J. H. (1973) *Black Power/White Control: The Struggle of the Woodlawn Organization in Chicago.* Princeton, NJ: Princeton University Press.

FISHBEIN, A. J. and ZINSMEYER, J. (1980) *Neighborhood-Based Reinvestment Strategies: A CRA Handbook.* Washington: U.S. Department of Housing and Urban Development.

FISHER, R. (1981) "From grass-roots organizing to community service: community organization practice in the community center movement, 1907-1930," in R. Fisher and P. Romanofsky, *Community Organization for Institutional Change: A Historical Perspective.* Westport, CO: Greenwood Press.

FLANAGAN, J. (1981) *The Successful Volunteer Organization.* Chicago: Contemporary Books.

——— (1982) *The Grass Root Fundraising Book.* Chicago: Swallow Press.

FLEMING, D. and CONLEY, G. (1977) *An Analysis of the Service Base Mutiplier Effect on Employment in a Dayton, Ohio Neighborhood.* Kettering, OH: Charles F. Kettering Foundation.

FLETCHER, R. K. and FAWCETT, S. B. (1979) *The Skills Exchange.* Lawrence, KS: Institute of Public Affairs and Community Development, University of Kansas.

FOLEY, D. L. (1952) "Neighbors or urbanites," in *The University of Rochester Studies of Metropolitan Rochester.* Rochester: University of Rochester Press.

Ford Foundation (1973) *Community Development Corporations: A Strategy for Depressed Urban and Rural Areas.* New York: Ford Foundation.

FREDERICKSON, H. G. [ed.] (1972) "Curriculum essays on citizens, politics, and administration in urban neighborhoods." *Public Administration Review* 32, special issue (October).

——— [ed.] (1973) *Neighborhood Control in the 1970s: Politics, Administration and Citizen Participation.* New York: Chandler Publishing Co.

Fresno County Economic Opportunities Commission (1979) *Recycling in Your Community: A Guide to Making It Happen.* Fresno, CA: Author.

FRIED, M. (1963) "Grieving for a lost home," in L. J. Duhl (ed.) *The Urban Condition.* New York: Basic Books.

——— and GLEICHER, P. (1961) "Some sources of residential satisfaction in an urban slum." *Journal of the American Institute of Planners* 27, no. 4: 305-315.

FRIEDMAN, R. and SCHWEKE, W. [eds.] (1981) *Expanding the Opportunity to Produce: Revitalizing the American Economy through New Enterprise Development.* Washington: Corporation for Enterprise Development.

FROLAND, C., PANCOAST, D., CHAPMAN, N., and KIMBOKO, P. (1979) *Professional Partnerships with Informal Helpers.* Presented to the American Psychological Association in New York (September).

——— (1981) *Helping Networks and Human Services.* Beverly Hills, CA: Sage Publications.

FULLER, J. (1977) *Municipal Advisory Councils: An Experiment in Community Participation.* Sacramento, CA: Office of Planning and Research, State of California.

GALLION, A. B. and EISNER, S. (1975) *The Urban Pattern: City Planning and Design.* New York: Van Nostrand.

GANDHI, M. K. (1934) *Hind Swaraj.* (Indian Home Rule). Madras, India: G.A Nateson & Co.

GANS, H. J. (1962a) "Urbanism and suburbanism as ways of life; a re-evaluation of definition," in A.M. Rose (ed.) *Human Behavior and Social Processes; An Interactionist Approach.* Boston: Houghton Mifflin Co.

——— (1962b) *The Urban Villagers.* New York: Free Press.

——— (1967) *The Levittowners.* New York: Pantheon.

——— (1974) *More Equality.* New York; Random House.

GARN, H. A., TEVIS, N. L., and SMEAD, C. E. (1976) *Evaluating Community Development Corporations.* Washington: Urban Institute.

GARTNER, A. and REISSMAN, F. (1980) *Self-Help in Human Services.* San Francisco: Jossey-Bass.

GITTELL, M. (1980) *Limits to Citizen Particiation.* Beverly Hills, CA: Sage Publications.

GLASS, T. E., and SANDERS, W. D. (1978) *Community Control in Education.* Midland, MI: Pendell Publishing Co.

GLUECK, E. T. (1927) *Community Schools.* Baltimore: Williams & Wilkins.

GOERING, J. M. (1979) "National neighborhood movement: a preliminary analysis and critique." *Journal of American Planning Association* 45: 506-514.

GOETZE, R. (1976) *Building Neighborhood Confidence: A Humanistic Strategy.* Cambridge, MA: Ballinger.

——— (1979) *Understanding Neighborhood Change: The Role of Expectations in Urban Revitalization.* Cambridge, MA: Ballinger.

——— (1980) *Neighborhood Monitoring and Analysis: A New Way of Looking at Urban Neighborhoods and How They Change.* Washington: Office of Policy Development and Research, U.S. Department of Housing and Urban Development.

GOLDSTEIN, B. and DAVIS, R. [eds.] (1977) *Neighborhoods in the Urban Economy.* Lexington, MA: Lexington Books.

GOTTLIEB, B. H. [ed.] (1981) *Social Networks and Social Support.* Beverly Hills, CA: Sage Publications.

GOURASH, H. (1978) "Help seeking: a review of the literature." *American Journal of Community Psychology* 6: 413-425.

GREELEY, A. M. (1977) *Neighborhood.* New York: Seabury Press.

GREENBAUM, P. E. and GREENBAUM, S. D. (1981) "Territorial personalization: group identity and social interaction in a Slavic-American neighborhood." *Environment and Behavior* 13, no. 5 (September): 574-589.

GREENBERG, St. (1974) *Politics and Poverty: Modernization and Response in Five Poor Neighborhoods.* New York: John Wiley.

GREENSTONE, J. D. and PETERSON, P. E. (1973) *Race and Authority: Community Participation in the War on Poverty.* New York: Russel Sage Foundation.

GREER, S. A. (1956) "Urbanism reconsidered: a comparative study of local areas in a metropolis." *American Sociological Review* 21 (February):

——— (1962) *The Emerging City: Myth and Reality.* New York: Free Press.

——— and GREER, A. L. [eds.] (1974) *Neighborhood and Ghetto: The Local Area in Large Scale Society.* New York: Basic Books.

GRIFFIN, B. W. (1981) *Cities within a City: On Changing Cleveland's Government.* Cleveland: College of Urban Affairs, Cleveland State University.

GRIGSBY, W. G. (1963) *Housing Markets and Housing Policy.* Philadelphia: University of Pennsylvania Press.

GUSFIELD, J. R. (1975) *Community: A Critical Response.* New York: Harper & Row.

GUTERMAN, S. S. (1969) "In defense of Wirth's 'urbanism as a way of life.' " *American Journal of Sociology* 74: 492-499.

GUTMAN, R. and POPENOE, D. [eds.] (1970) *Neighborhood, City and Metropolis: An Integrated Reader in Urban Sociology.* New York: Random House.

HALLETT, S. J. (1978) *The Neighborhood Economy: Frameworks for the Flow of Funds.* Evanston, ILL: Center for Urban Affairs, Northwestern University.

HALLMAN, H. W. [ed.] (1958) *Forum on Neighborhoods Today & Tomorrow* (6 issues). Philadelphia: Philadelphia Housing Association.

——— (1959) "Citizens and professionals reconsider the neighborhood." *Journal of the American Institute of Planners* 25, no. 3 (August): 121-127.

——— (1967) "The community action program — an interpretative analysis of 35 communities" in *Examination of the War on Poverty: Staff and Consultants Reports, Vol. 4.* Prepared for U.S. Senate Subcommittee on Employment, Manpower, and Poverty. Washington: Government Printing Office. Reprinted in W. Bloomberg, Jr., and H. J. Schmandt (ed.) *Power, Poverty and Urban Policy* Beverly Hills, CA: Sage Publications.

——— (1970) *Neighborhood Control of Public Programs: Case Studies of Community Corporations and Neighborhood Boards. New York: Praeger.*

——— (1974) *Neighborhood Government in a Metropolitan Setting.* Beverly Hills, CA: Sage Publications.

——— (1977a) *The Organization and Operation of Neighborhood Councils: A Practical Guide.* New York: Praeger.

——— (1977b) *Small and Large Together: Governing Metropolitan Areas.* Beverly Hills, CA: Sage Publications.

——— (1978) *Citizen Involvement in the Local Budget Process.* Washington: Center for Community Change.

——— (1980) *Community-Based Employment Programs.* Baltimore: Johns Hopkins University Press.

——— (1981a) *Citizens and Program Implementation.* Washington: Civic Action Institute.

——— (1981b) *Innovative Citizen Roles in Local Budget Making.* Washington: Civic Action Institute.

——— (1981c) *Neighborhood Planning.* Washington: Civic Action Institute.

——— (1981d) "Patterns of CDBG citizen participation: a survey." *Neighborhood Ideas* 6, no. 1 (September): 1, 12-15.

——— (1981e) "Neighborhood roles in governance." *Neighborhood Ideas* 6, no. 2 (October/November): 17, 20-23.

——— (1983) *Neighborhood Governance.* Washington: Office of Policy Development and Research, U.S. Department of Housing and Urban Development.

——— and GOLDWASSER, T. (1980) *What Neighborhoods Can Do about the Energy Crisis.* Washington: Civic Action Institute.

——— and WEGENER, E. (1980) *Community Organizing in a Partnership Setting.* Washington: Civic Action Institute.

HAMBLETON, R. (1979) *Policy Planning and Local Government.* Montclair, NJ: Allanheld, Osmun.

HANNERZ, U. (1969) *Soulside: Inquiries into Ghetto Culture and Community.* New York: Columbia University Press.

——— (1980) *Exploring the City: Inquiries Toward an Urban Anthropology.* New York: Columbia University Press.

HANSON, R. and McNAMARA, J. (1981) *Partners.* Minneapolis: Dayton Hudson Foundation.

HARRIS, C. D. and ULMAN, E. L. (1945) "The nature of cities." *Annals of the American Academy of Political and Social Science* 242 (November): 7-17.

HARTMAN, C., KEATING, D., and LeGATES, R. (1982) *Displacement: How to Fight It.* Berkeley, CA: National Housing Law Project.

HATRY, H. P. and VALENTE, C. F. (1983) "Alternative service delivery approaches involving increased use of the private sector" in *The Municipal Yearbook, 1983.* Washington: International City Management Association.

HAWKINS, R. B., Jr. (1976) *Self-Government by District.* Stanford, CA: Hoover Institution Press.

HAWORTH, L. (1963) *The Good City.* Bloomington, IN: Indiana University Press.

HENDERSON, A. T. (1981) *Parent Participation -- Student Achievement: The Evidence Grows.* Columbia, MD: National Committee for Citizens in Education.

HENIG, J. R. (1982) *Neighborhood Mobilization: Redevelopment and Response.* New Brunswick: Rutgers University Press.

HENZE, L., KIRSHNER, K, and LILLOW, L. (1979) *An Income and Capital Flow Study of East Oakland, California.* Oakland: Community Economics.

HERBERT, G. (1963) "The neighborhood unit principle and organic theory" in *The Sociological Review* 11 (July): 165-213.

HERLIG, P. and MUNDT, R. J. (1982) "Districts and city council decision-making" in *Urban Affairs Quarterly* 17, no. 3 (March): 371-377.

HESKINS, A. D. (1980) "Crisis and response: a historical perspective on advocacy planning" in *Journal of the American Planning Association* 46 (January): 50-63.

HESTER, R. T. (1975) *Neighborhood Space.* Stroudsburg, PA: Dowden, Hutchinson, and Ross.

HILL, R. B. (1972) *The Strengths of Black Families.* New York: National Urban League.

HILLERY, G. A., Jr. (1955) "Definitions of community: areas of agreement." *Rural Sociology* 20 (June): 11.

HILLMAN, A. (1960) *Neighborhood Centers Today.* New York: National Federation of Settlements and Neighborhood Centers.

HINCKLEY, K. A. (1977) "The bang and the whimper: model cities and ghetto opinion." *Urban Affairs Quarterly* 13, no. 2 (December): 131-150.

HIPPLER, A. E. (1974) *Hunter's Point: A Black Ghetto.* New York: Basic Books.

HOLLISTER, R. M., KRAMER, B. M., and BELLIN, S. S. (1974) *Neighborhood Health Centers.* Lexington, MA: Lexington Books.

HOOVER, E. M. and VERNON, R. (1959) *Anatomy of a Metropolis: The Changing Distribution of People and Jobs within the New York Metropolitan Region.* Cambridge, MA: Harvard University Press.

Housing Action Council (1980) *Community Revitalization Bibliography.* Washington: U.S. Department of Housing and Urban Development.

HOWARD, E (1965) *Garden Cities of Tomorrow* (1902). Cambridge, MA: MIT Press.

HOYT, H. (1939) *The Structure and Growth of Residential Neighborhoods in American Cities.* Washington: Federal Housing Administration.

HUGHES, J. W. and BLEAKLY, K. D., Jr. (1975) *Urban Homesteading.* New Brunswick, NJ: Center for Urban Policy Research, Rutgers University.

HUNTER, A. (1974) *Symbolic Communities: The Persistence of Change in Chicago's Communities.* Chicago: University of Chicago Press.

——— (1975) "The loss of community: an empirical test through replication." *American Sociological Review* 40 (October): 537-552.

——— (1979) "The urban neighborhood: its analytical and social context." *Urban Affairs Quarterly* 14, no. 3 (March): 267-288.

Institute for Community Economics (1982) *The Community Land Trust Handbook.* Emmaus, PA: Rodale Press.

Institute for Local Self-Reliance (1980) "Methods for measuring community cash flow." *Self-Reliance* 23 (July/August): 1, 4-5, 11.

Institute for Responsive Education (1980) "Making Parent Councils Effective." in *Citizens Action in Education* 7, no. 1.

International City Management Association [ICMA] (1977a) "Using citizen surveys: three approaches." *Innovations* 15. Washington: Author.

——— (1977b) "Community relations." *MIS Report* 9, no. 5 (May). Washington: Author.

——— (1981) "Residential fire protection." *MIS Report* 13, no. 1 (January). Washington: Author.

ISAACS, R. (1948a) "The neighborhood theory." *Journal of the American Institute of Planners* 14 (Spring): 15-23.

——— (1948b) "Are urban neighborhoods possible?" *Journal of Housing* 5, no. 7 (July): 177-180.

——— (1948c) "The neighborhood unit is an instrument for segregation." *Journal of Housing* 5, no. 8 (August): 215-218.

JACOBS, J. (1961) *The Death and Life of Great American Cities.* New York: Random House.

JACOBSON, S. (1983) *Neighborhood Control of Health and Social Service Delivery: A Concept Paper.* Washington: Office of Policy Development and Research, U.S. Department of Housing and Urban Development.

JANOWITZ, M. (1967), *The Community Press in an Urban Setting: The Social Elements of Urbanism* (1952). Chicago: University of Chicago Press.

——— and SUTTLES, G. (1978) "The social ecology of citizenship" in R. Sarri and Y. Hasenfelds (ed.) *The Management of Human Services.* New York: Columbia University Press.

JEFFERSON, T. (1943) *The Complete Jefferson* (arranged by P.W. Padover). New York: Duell, Sloan & Pearce.

JOHNSEN, R., TOBIN, D., and BOND, J. (1982) *Organizing for Local Fundraising.* Boulder, CO: VOLUNTEER, the Center for Citizen Involvement.

JOHNSTON, R. J. (1971) *Urban Residential Patterns.* New York: Praeger.

JONES, V. (1942) *Metropolitan Government.* Chicago: University of Chicago Press.

JORDAN, D. et al. (1976) *Effective Citizen Participation in Transportation Planning.* 2 vol. Washington: U.S. Department of Transportation.

KAHN, S. (1982) *Organizing: A Guide for Grassroots Leaders.* New York: McGraw-Hill.

KAIN, J. K. and APGAR, W. C., Jr. (1979) "Modeling neighborhood change" in D. Segal, ed. *The Economics of Neighborhoods.* New York: Academic Press.

KAPLAN, H. (1963) *Urban Renewal Politics.* New York: Columbia University Press.

KAPLAN, M. (1969) "Advocacy and the Urban Poor" in *Journal of the American Institute of Planners* 35, no. 2 (March): 96-101.

KARNIG, A. K. (1976) "Black representation in city council: the impact of district elections and socioeconomic factors" and "A response to Cole" in *Urban Affairs Quarterly* 12, no. 2 (December): 223-242; 251-256.

KASARDA, J. D. and JANOWITZ, M. (1974) "Community attachment in mass society." *American Sociological Review* 39 (June): 328-399.

KASPERSON, R. E. and BREITBART, M. (1974) *Participation, Decentralization and Advocacy Planning.* Washington: Association of American Geographers.

KATZNELSON, I. (1981) *City Trenches: Urban Politics and the Patterning of Class in the U.S.* New York: Pantheon.

KELLER, S. (1968) *The Urban Neighborhood: A Sociological Perspective.* New York: Random House.

KELLY, R. M. (1977) *Community Conrol of Economic Development: The Boards of Directors of Community Development Corporations.* New York: Praeger.

KEYES, L. C., Jr. (1969) *The Rehabilitation Planning Game.* Cambridge, MA: MIT Press.

KING, M. L., Jr. (1963) *Strength to Love.* New York: Harper & Row.

KLEIN, J. F. (1980) *Organizing for Neighborhood Justice.* Chicago: Center for Urban Policy, Loyola University of Chicago.

KOBRIN, S. (1959) "The Chicago area project — a 25 year assessment." *Annals of the American Academy of Political and Social Science* 322 (March): 19-27.

KOLLIAS, K. (1977) *Neighborhood Reinvestment: A Citizen's Compendium for Programs and Strategies.* Washington: National Center for Urban Ethnic Affairs.

KORNBLUM, W. (1974) *Blue Collar Community.* Chicago: University of Chicago Press.

KOTLER, M. (1969) *Neighborhood Government: The Local Foundations of Political Life.* Indianapolis: Bobbs-Merrill.

——— (1981) *Report and Recommendations on Neighborhood Service Delivery.* Washington: Center for Responsive Governance.

——— (1982) "Partnerships in community service." *Journal of Community Action* 1, no. 4: 45-50.

——— (1983) *Neighborhood Delivery of Environmental Services.* Washington: Office of Policy Development and Research, U.S. Department of Housing and Urban Development.

KOTLER, N. (1978) *Neighborhood Economic Enterprises.* Kettering, OH: Charles F. Kettering Foundation.

KRAMER, R. M. (1969) *Participation of the Poor: Comparative Community Case Studies in the War on Poverty.* Englewood Cliffs, NJ: Prentice-Hall.

KRISTOL, I. (1968) "Decentralization for what?" *The Public Interest* (Spring): 17-25.

KULKA, R., VEROFF, J., and DOUVAN, E. (1979) "Social class and the use of professional help for personal problems: 1957 and 1976." *Journal of Health and Social Behavior* 20: 2-17.

KWEIT, R. W. and KEIL, M. G. (1980) "Bureaucratic decision making: impediments to citizen participation." *Polity* 12: 647-666.

LACHMAN, M.L. and DOWNS, A. (1978) "The role of neighborhoods in the mature metropolis," in C.L. Leven (ed.) *The Mature Metropolis.* Lexington, MA: Lexington Books.

LAMB, C. (1975) *Political Power in Poor Neighborhoods.* New York: Halstead Press.

LANCOURT, J. E. (1979) *Confront or Concede: The Alinsky Citizen-Action Organizations.* Lexington, MA: D. C. Heath.

LANE, R. (1967) *Policing the City: Boston, 1822-1885.* Cambridge, MA: Harvard University Press.

LANG, M. H. (1982) *Gentrification Amid Urban Decline.* Cambridge, MA: Ballinger.

LANGTON, S. [ed.] (1978) *Citizen Participation in America: Essays on the State of the Art.* Lexington, MA: Lexington Books.

——— [ed] (1979) *Citizen Participation: Proceedings of the National Conference on Citizen Participation.* Medford, MA: Lincoln Filene Center for Citizenship and Public Affairs, Tufts University.

LA NOUE, G. R. and SMITH, L. R. (1973) *The Politics of School Decentralization.* Lexington, MA: Lexington Books.

LASKA, S. B. and SPAIN, [eds.] (1980) *Back to the City.* New York: Pergamon Press.

Lawrence Johnson & Associates (1978) *Citizen Participation in Community Development: A Catalog of Local Approaches.* Washington: Government Printing Office.

LEACH, R. H. (1970) *American Federalism.* New York: Norton.

League of Women Voters of the United States (1974) *The Citizen and the Budget Process: Opening Up the System.* Washington: Author.

LEINHARDT, S. [ed.] (1977) *Social Networks: A Developing Paradigm.* New York: Academic Press.

LEVEN, C. L., LEVEN, J. L., MOURSE, H. O., and READ, R. B. (1976) *Neighborhood Change: Lessons in the Dynamics of Urban Decay.* New York: Praeger.

——— [ed.] (1978) *The Mature Metropolis.* Lexington, MA: Lexington Books.

LEVIN, H. M. [ed.] (1970) *Community Control of Schools.* Washington: Brookings Institute.

LEVITAN, S. A. (1969) *The Great Society's Poor Law: A New Approach to Poverty.* Baltimore: Johns Hopkins Press.

——— and MANGUM, G. L. (1969) *Federal Training and Work Programs in the Sixties.* Ann Arbor, MI: Institute of Labor and Industrial Relations.

LEVITAS, G. (1978) "Anthropology and sociology of streets" in S. Anderson, ed., *On Streets.* Cambridge, MA: MIT Press.

LEVY, F. S., MELTSNER, A. J., and WILDAVSKY, A. (1974) *Urban Outcomes: Schools, Streets, and Libraries.* Berkeley, CA: University of California Press.

LEWIS, G. (1959) "Citizen participation in renewal surveyed." *Journal of Housing* 16, no. 3 (March): 80-89.

LIEBERMAN, M. and MULLAN, J. (1978) "Does help help: the adaptive consequences of obtaining help from professionals and social networks." *American Journal of Community Psychology* 6: 499-517.

LIEBOW, E. (1967) *Tally's Corner.* Boston: Little, Brown.

LIPSET, S. M. (1963) *Political Man: The Social Basis of Politics.* Garden City, NY: Doubleday.

LITWAK, E. (1978) "Agency and family linkages in providing neighborhood services," in D. Thurz and J.L. Vigilante (eds.) *Reaching People: The Structure of Neighborhood Services.* Beverly Hills, CA: Sage Publications.

——— and SZELENYI, I. (1969) "Primary group structures and their functions." *American Sociological Review* 34 (August): 465-481

LITWAK, L. and DANIELS, B. (1979) *Innovations in Development Financing.* Washington: Council of State Planning Agencies.

LOCKE, H. G. (1977) "The evolution of contemporary police services," in B. L. Garmire (ed.) *Local Government Police Management.* Washington: International City Management Association.

LOCKE, J. (1960) *Two Treatises of Government.* New York: New American Library.

LONDON, B., BRADLEY, D. S, and HUDSON, J. R. (1980) "The revitalization of inner-city neighborhoods" [a symposium]. *Urban Affairs Quarterly* 15 (June): 373-487.

LONGHINI, G. and MOSENCE, D. (1978) *Homeowners' Associations: Problems and Remedies.* Chicago: American Planning Association.

LORING, W. C., Jr., SWEETSER, F. L., and ERNST, C. F. (1957) *Community Organization for Citizen Participation in Urban Renewal.* Boston: Massachusetts Department of Commerce.

LYNCH, K. (1960) *The Image of the City.* Cambridge, MA.: Technology Press and Harvard University Press.

——— (1981) *A Theory of Good City Form.* Cambridge, MA: MIT Press.

McCLAUGHERY, J. (1980) "Neighborhood revitalization," in P. Duignan and A. Ravushka (ed.) *The United States in the 1980s.* Stanford, CA: Hoover Institution Press.

McDonough, Bond & Associates (1983) *Neighborhood Fiscal Empowerment.* Washington: Office of Policy Development and Research, U.S. Office of Housing and Urban Development.

McGAHAN, P. (1972) "The neighbor role and neighboring in a highly urban area." *Sociological Inquiry* 13 (Summer): 397-408.

McGILLIS, D. and McGILLIS, J. (1977) *Neighborhood Justice Centers: An Analysis of Potential Models.* Washington: National Institute of Law Enforcement and Criminal Justice, U.S. Department of Justice.

McKENSIE, R. (1923) *The Neighborhood.* Chicago: University of Chicago Press.

McNULTY, R. H. and KLIMENT, S. A. [ed.] (1976) *Neighborhood Conservation: A Handbook of Methods and Techniques.* New York: Whitney Library of Design.

MADISON, J. (n.d.) "The Federalist No. 46," in *The Federalist.* Modern Library Edition. New York: Random House.

MAHMOOD, S. T. and GHOSH, A. (1979) *Handbook for Community Economic Development.* Washington: Government Printing Office.

MANN, P. (1970) "The Neighborhood," in R. Gutman and D. Popenoe (ed.) *Neighborhood, City, and Metropolis.* New York: Random House.

MANN, S. [ed.] (1969) *Proceedings of National Conference on Advocacy and Pluralistic Planning.* New York: Urban Research Center, Hunter College of the City University of New York.

MARCINIAK, E. (1977) *Reviving an Inner City Community.* Chicago: Loyola University of Chicago.

——— (1981) *Reversing Urban Decline.* Washington: National Center for Urban Ethnic Affairs.

MARR, P. D. [ed.] (1973) *Multi-Service Centers: Innovations in the Delivery of Welfare Services.* Davis, CA: Institute of Governmental Affairs, University of California, Davis.

MARRIS, P. and REIN, M. (1967) *Dilemmas of Social Reform: Poverty and Community Action in the United States.* London: Routledge & Kegan Paul.

MARSHALL, D. R. (1971) *The Politics of Particiation in Poverty.* Berkeley: University of California Press.

Marshall Kaplan, Gans & Kahn (1973a) *The Model Cities Program: A Comparative Analysis of Particiating Cities Process, Product, Performance, and Prediction.* Washington: U.S. Department of Housing and Urban Development.

——— (1973b) *Model Cities Program: A Comparative Analysis of City Response Patterns and Their Relation to Future Urban Policy.* Washington: U.S. Department of Housing and Urban Development.

MARSHALL, P. [ed.] (1977) *Citizen Participation for Community Development: A Reader on the Citizen Participation Process.* Washington: National Association of Housing and Redevelopment Officials.

MARSHALL, S. A. and Mayer, N. (1983) *Neighborhood Organizations and Community Development.* Washington: Urban Institute.

MARTINEAU, W. H. (1976) "Social participation and a sense of powerlessness among blacks: a neighborhood analysis." *Sociological Quarterly* 17 (Winter): 27-41.

——— (1977) "Informal ties among urban black Americans: some new data and a review of the problems." *Journal of Black Studies* 8, no. 1 (September): 83-104.

MAUER, R. and BAXTER, J. (1972) "Images of the neighborhood and city among black, Anglo, and Mexican-American children." *Environment and Behavior* 4: 351-388.

MAYER, N. S. with J. L. BLAKE (1981) *Keys to the Growth of Neighborhood Development Organizations.* Washington: Urban Institute Press.

MAYO, J. M., Jr. (1979) "Effects of street forms on suburban neighboring behavior." *Environment and Behavior* 11, no. 3 (September): 375-397.

MAZZIOTTI, D. F. (1974) "The underlying assumptions of advocacy planning: pluralism and reform." *Journal of the American Institute of Planners* 40, no. 1 (January): 38-47.

MERRY, S. E. (1981a) "Defensible space undefended: social factors in crime control through environmental design." *Urban Affairs Quarterly* 16, no. 4 (June): 397-422.

——— (1981b) *Urban Danger.* Philadelphia: Temple University Press.

MEYER, J. A. (1982) "Private-sector initiatives and public policy: a new agenda," in J. A. Meyer, (ed.) *Meeting Human Needs.* Washington: American Enterprise Institute.

MICHENER, J. (1978) *The Back to the City Movement: The Possibilities of Increasing Racial and Economic Integration.* Washington: National Neighbors, Inc.

MILBRAITH, L. W. and GOEL, M. L. (1977) *Political Participation: How and Why Do People Get Involved in Politics?* Chicago: Rand McNally.

MILGRAM, M. (1977) *Good Neighborhood: The Challenge of Open Housing.* New York: Norton.

MILLAS, A. J. (1980) "Planning for the elderly within the context of a neighborhood." *Ekistics* 283 (July/August): 264-272.

MILLER, M. (1979) *The "Ideology" of the Community Organization Movement.* San Francisco: Organize Training Center.

MILLMAN, H. (1960) "Jewish community centers," in *American Jewish Yearbook*, vol. 61: 92. New York: American Jewish Committee.

MILLSPAUGH, M. and BRECKENFELD, G. (1958) *The Human Side of Urban Renewal: A Study of Attitude Changes Produced by Neighborhood Rehabilitation.* Baltimore: Fight Blight, Inc.

MINZEY, J. D. (1981) "Community education and community schools," in D. Davies, (ed.) *Communities and Their Schools.* New York: McGraw-Hill.

MIRENGOFF, W., RINDLER, L., GREENSPAN, H., and SEABLOM, S (1979) *CETA: Assessment of Public Service Employment Programs.* Washington: National Academy of Sciences.

MIRENGOFF, W., RINDLER, L., GREENSPAN, H., and HARRIS, C. (1982) *CETA: Accomplishments, Problems, Solutions.* Kalamazoo, MI: W. E. Upjohn Institute for Employment Research.

MITCHELL, J. C. [ed.] (1969) *Social Networks in Urban Situations: Analysis of Personal Relationships in Central African Towns.* Manchester, England: Manchester University Press.

Model Cities Service Center (1971) *Model Cities: A Report on Progress..* Washington: National League of Cities/U.S. Conference of Mayors.

MOGULOFF, M. (1970) *Citizen Participation: The Local Perspective.* Washington: Urban Institute.

MOLES, O. (1981) "Home-school partnerships." in *Citizen Action in Education* 8, no. 2 (November).

——— COLLINS, C., and CROSS, M. (1982) *The Home-School Connection: Selected Partnership Programs in Large Cities.* Boston: Institute for Responsive Education.

MORRIS, D. and HESS, K. (1975) *Neighborhood Power: The New Localism.* Boston: Beacon Press.

MUDD, J. (forthcoming) book on New York decentralization experience. New Haven, CT: Yale University Press.

MULLIGAN, K. and POWELL, J. (1979) *Operating a Recycling Program: A Citizen's Guide.* Washington: U.S. Environmental Protection Agency.

MUMFORD, L. (1961) *The City in History.* New York: Harcourt Brace Jovanovitch.

——— (1968) *The Urban Prospect.* New York: Harcourt Brace Jovanovitch.

MURPHY, C. G. (1960) "Community organization for social welfare," in R. H. Kurtz (ed.), *Social Work Year Book, 1960.* New York: National Association of Social Workers.

MURPHY, R. D. (1971) *Political Entrepreneurs and Urban Poverty: The Strategies of Policy Innovation in New Haven's Model Anti-Proverty Project.* Lexington, MA: D. C. Health and Lexington Books.

NAPARSTEK, A. J. (1976) "Policy options for neighborhood empowerment," in *Urban Options I.* Columbus, OH: Academy for Contemporary Problems.

——— and CINCOTTA, G. (1976) *Urban Disinvestment: New Implications for Community Organization Research and Public Policy.* Washington and Chicago: National Center for Urban Ethnic Affairs and National Training and Information Center.

NAPARSTEK, A. J., BIEGEL, D. E., and SPIRO, H. R. (1982) *Neighborhood Networks for Human Mental Health Care.* New York: Plenum.

NATHAN, R. P. et al. (1978) *Monitoring the Public Service Employment Program.* Washington: National Commission for Manpower Policy.

NATHAN, R. P., COOK, R. F., RAWLINS, V. L, and associates (1981) *Public Service Employment: A Field Evaluation.* Washington: Brookings Institution.

National Academy of Public Administration [NAPA] (1977) *Multi-tiered Metropolitan Government: Four U.S. Reform Efforts.* Washington: Author.

National Advisory Commission on Civil Disorders (1968) *Commission Report.* New York: Bantam Books.

National Advisory Council on Economic Opportunity [NACEO] (1968) *Decentralization to Neighborhoods: A Conceptual Analysis.* Washington: Government Printing Office.

National Association of Housing and Redevelopment Officials [NAHRO] (1979) *Designing Rehabilitation Programs: A Local Government Guidebook.* Washington: Government Printing Office.

National Association of Neighborhoods (1978) *NAN Handbook of State and Local Neighborhood Legislation.* Washington: Author.

National Center for Economic Alternatives (1981) *Federal Assistance to Community Development Corporations: An Evaluation of Title VII of Community Services Act of 1974.* Washington: Author.

National Center for Urban Ethnic Affairs [NCUEA] (1979a) *Community Development Credit Unions.* Washington: Author.

——— (1979b) *Neighborhood Commercial Revitalization: A Workshop Report.* Kansas City, Missouri, August 13-14, 1979. Washington: Author.

National Commission on Neighborhoods (1979) *Neighborhoods, People, Building Neighborhoods.* With a two volume appendix. Washington: Government Printing Office.

National Commission on Urban Problems (1968) *Building the American City.* Washington: Government Printing Office.

National Council for Urban Economic Development [NCUED] (1979) "Neighborhood commercial revitalization." *Information Service* 17 (March).

National Development Council [NDC] (c.1979) *Neighborhood Business Revitalization.* Washington: Author.

National Economic Development and Law Center (1983) *Community Economic Development Strategies* (3 vol.). Berkeley, CA: Author.

National Peoples Action (1975) *The Grass-Roots Battle Against Redlining: From the Streets to the Halls of Congress.* Chicago: Author.

National Research Council, Social Science Panel on the Significance of Community in the Metropolitan Environment (1974) *Toward an Understanding of Metropolitan America.* San Francisco: Canfield Press.

National Trust for Historic Preservation (1979) "Commercial Revitalization: A Conservation Approach," in *Conserve Neighborhoods* 7 (Summer).

——— (1980) "Community Events: How to Organize Them," in *Conserve Neighborhoods* 13.

——— *Conserve Neighborhoods.* (A bimonthly bulletin). Washington, D.C.

National Urban Coalition (1978) *Displacement: City Neighborhoods in Transition.* Washington: Author.

National Urban Development Services Corporation (c. 1977) *Neighborhood Commercial Revitalization.* Washington: Author.

National Urban League (c. 1973) *Toward Effective Citizen Participation in Urban Renewal.* New York: Author.

NEEDLEMAN, M. L. and NEEDLEMAN, C. E. (1974) *Guerrillas in the Bureaucracy: The Community Planning Experience in the United States.* New York: John Wiley.

NEWMAN, O. (1972) *Defensible Space: Crime Prevention through Urban Design.* New York: Macmillan.

——— (1980) *Community of Interest.* Garden City, NY: Doubleday.

——— and FRANCK, K. A. (1980) *Factors Influencing Crime and Instability in Urban Housing Developments.* Washington: Government Printing Office.

NISBET, R. (1975) *Twilight of Authority.* New York: Oxford University Press.

NOHARA, S. (1968) "Social context and neighborliness: the Negro in St. Louis," in S. Greer et al. (eds.) *The New Urbanization* New York: St. Martin's.

NORLINGER, E. A. (1972) *Decentralizing the City: A Study of Boston's Little City Halls.* Boston: Boston Urban Observatory.

O'BRIEN, D. J. (1975) *Neighborhood Organization and Interest Group Process.* Princeton, NJ: Princeton University Press.

Ohio Commission on Local Government Services (1974) *Neighborhood Governance.* Columbus, OH: Author.

OLSON, M. (1965) *The Logic of Collection Action: Public Goods and the Theory of Groups.* Cambridge, MA: Harvard University Press.

OLSON, P. (1982) "Urban neighborhood research: its development and current focus" in *Urban Affairs Quarterly* 17, no. 4 (June): 491-518.

"Organizing neighborhoods" (1979) *Social Policy* 10, no. 2 (September/October): entire issue.

ORNSTEIN, A. C. (1974) *Metropolitan Schools: Administrative Decentralization vs. Community Control.* Metuchen, NJ: Scarecrow Press.

PALMER, J. L. and SAWHILL, I. V. [eds.] (1982) *The Reagan Experiment.* Washington: Urban Institute Press.

PARENTI, M. (1970) "Power and pluralism: a view from the bottom." *Journal of Politics* 32 (August): 501-530.

PATEMAN, C. (1970) *Participation and Democratic Theory.* Cambridge, England: Cambridge University Press.

PEATTIE, L. R. (1968) "Reflections on advocacy planning." *Journal of American Institute of Planners* 32, no. 2 (March): 80-88.

PERLMAN, J. (1976) "Grassrooting the system." *Social Policy* 7, no. 2 (September/ October): 4-20.

——— (1979) "Grassroot empowerment and government response." *Social Policy* 10, no. 2 (September/October): 16-21.

PERRY, C. (1929) "The neighborhood unit." *Neighborhood and Community Planning, vol. 7, Regional Plan of New York and Its Environs.* New York: Russell Sage Foundation.

PERRY, D. C. (1973) "The Suburb as a Model for Neighborhood Control," in G. Frederickson (ed.) *Neighborhood Control in the 1970s.* New York: Chandler Publishing Co.

PIVEN, F. F. and CLOWARD, R. A. (1971) *Regulating the Poor: The Functions of Public Welfare.* New York: Random House.

——— (1982) *The New Class War: Reagan's Attack on the Welfare State and Its Consequences.* New York: Pantheon.

POLLINGER, K. J. and POLLINGER, A. C. (1972) *Community Action and the Poor: Influence vs. Social Control in a New York City Community.* New York: Praeger.

POWLEDGE, F. (1970) *Model City: A Test of American Liberalism: One Town's Efforts to Rebuild Itself.* New York: Simon & Schuster.

President's Commission on Mental Health (1978) *Report to the President (4 vol.).* Washington: Government Printing Office.

President's Task Force on Private Sector Initiatives (1982) *Building Partnerships.* Washington: Author.

Public Affairs Counseling (1975) *The Dynamics of Neighborhood Change.* Washington: U.S. Department of Housing and Urban Development.

RAIN (1977) *Rainbook.* Portland, OR: RAIN.

RAINWATER, L. (1966) "Fear and the house-as-haven in the lower class." *Journal of the American Institute of Planners* 32, no. 1 (January): 23-31

——— (1970) *Behind Ghetto Walls: Black Families in a Federal Slum.* Chicago: Aldine.

RAMATI, R. (1981) *How to Save Your Own Street.* Garden City, NY: Doubleday.

REES, I. B. (1971) *Goverment by Community.* London: Charles Knight & Co.

RICH, L.G., NETHERWOOD, J. C., and CAHN, E. G. (1981) *Neighborhood: A State of Mind.* Baltimore: Johns Hopkins University Press.

RICH, R. C. (1977) "Equity and institutional design in urban service delivery." *Urban Affairs Quarterly* 12, no. 3 (March): 383-410.

——— (1980a) "The dynamics of leadership in neighborhood organizations." *Social Science Quarterly* 60, no. 4 (March): 570-587.

——— (1980b) "A political economy approach to the study of neighborhood organizations." *American Journal of Political Science* 24 (November): 559-592.

——— (1981) "The interaction of the voluntary and governmental sectors: toward an understanding of coproduction of municipal services." *Administration and Society* 13 (May): 59-76.

——— (forthcoming) "Balancing autonomy and capacity in community organizations."

——— and ROSENBAUM, W.A. [eds.] (1981) "Citizen participation in public policy" [a symposium]. *Journal of Applied Behavioral Science* 17, no. 4: 436-614.

ROBINSON, T. P. and DYE, T. D. (1978) "Reformism and black representation on city councils." *Social Science Quarterly* 59 (June): 133-141.

ROGERS, D. (1982) "School decentralization: it works." *Social Policy* 12 (Spring): 13-23.

ROHE, W. and GATES, L. B. (1983) forthcoming article on neighborhood councils in *Journal of Community Action* 2, no. 1.

ROSE, S. M. (1972) *The Betrayal of the Poor: The Transformation of Community Action.* Cambridge, MA: Schenkman Publishing Co.

ROSEN, G. (1978) "First neighborhood health center movement — its rise and fall," in J. W. Leavitt and R. L. Numbers (eds.) *Sickness and Health in America.* Madison, WI: University of Wisconsin Press.

ROSENBAUM, N. M. (1976) *Citizen Involvement in Land Use Governance: Issues and Methods.* Washington: Urban Institute.

——— [ed.] (c. 1980) *Citizen Participation: Models and Methods of Evaluation.* Washington: Center for Responsive Goverance.

——— and RICH, R. C. (1982) "Neighborhood councils in urban politics; patterns of citizen involvement in St. Paul." Presented at annual meeting of Midwest Political Science Association.

ROSENER, J. B. (1975) "A cafeteria of techniques and critiques." *Public Management* 57, no. 11 (December).

——— (1978) "Citizen participation: can we measure its effectiveness?" *Public Administration Review* 38 (September): 457-463.

——— (1981) "User-oriented evaluation: a new way to view citizen participation." *Journal of Applied Behavioral Science* 17, no. 4: 583-614.

ROSS, H. L. (1962) "The local community: a survey approach." *American Sociological Review* 27: 75-89.

ROSSI, P. H. and DENTLER, R. A. (1961) *The Politics of Urban Renewal: The Chicago Findings.* New York: Free Press.

ROTHMAN, J. (1974) *Planning and Organizing for Social Change: Action Principles from Social Science Research.* New York: Columbia University Press.

RUBIN, I. (1972) *Satmor: An Island in the City.* Chicago: Quadrangle Books.

RUSTIN, B. (1970) "The failure of black separatism." *Harper's Magazine* (January): 25-34.

SALEM, G. (1980) *The Forty-Fourth Ward Assembly.* Chicago: Center for Urban Policy, Loyola University of Chicago.

SANDLER, G. (1974) *The Neighborhood: The Story of Baltimore's Little Italy.* Baltimore: Bodine & Associates.

SAVAS, E. S. (1982) *Privatizing the Public Sector.* Chatham, NJ: Chatham House Publishers.

SCHAFER, R. and LADD, H. F. (1981) *Discrimination in Mortgage Lending.* Cambridge, MA: MIT Press.

SCHAFFER, R. L. (1973) *Income Flows in Urban Proverty Areas.* Lexington, MA: Lexington Books.

SCHMANDT, H. J. (1972) "Municipal decentralization: an overview" in *Public Administration Review* 32, special issue (October): 571-588.

——— (1973) "Decentralization: a structural imperative," in G. Frederickson (ed.) *Neighborhood Control in the 1970s.* New York: Chandler Publishing Co.

SCHOENBERG, S. P. (1979) "Criteria for the evaluation of neighborhood vitality in working class and low income areas in core cities" in *Social Problems* 27, no. 1 (October)

——— and ROSENBAUM, P. L. (1980) *Neighborhoods that Work: Sources for Vitality in the Inner City.* New Brunswick, NJ: Rutgers University Press.

SCHOR, L. B. and ENGLISH, J. T. (1974) "Background context and signficant issues in neighborhood health center program" in R. H. Hollister et al. (eds.) *Neighborhood Health Centers.* Lexington, MA: Lexington Books.

SCHULBERG, H. C. and KILLILEA, M. (1982) "Community mental health in transition" in H. C. Schulberg, and M. Killilea, (eds.) *Modern Practice of Community Mental Health.* San Francisco: Jossey-Bass.

SCHUMACHER, E. F. (1973) *Small Is Beautiful: Economics as if People Mattered.* New York: Harper & Row.

SCHUR, R. (1979) *Combating Housing Abandonment.* Washington: Civil Action Institute.

——— and SHERRY, V. (1977) *The Neighborhood Housing Movement.* New York: Association of Neighborhood Housing Developers.

SCHWARTZ, E. (1982) *The Neighborhood Agenda.* Philadelphia: Institute for the Study of Civic Values.

SEELEY, D. (1981) *Education through Partnership.* Cambridge, MA: Ballinger.

SEELEY, D. and SCHWARTZ, R. (1981) "Debureaucratizing public education: the experience of New York and Boston," in D. Davies (ed.) *Communities and Their Schools.* New York: McGraw-Hill.

SEGAL, D. [ed.] (1979) *The Economics of Neighborhoods.* New York: Academic Press.

SHALALA, D. E. (1971) *Neighborhood Governance: Issues and Proposals.* New York: American Jewish Committee.

SHERMAN, L. W., MILTON, C. H., and KELLY, T. V. (1973) *Team Policing: Seven Case Studies.* Washington: Police Foundation.

SHEVSKY, E. and WILLIAMS, M. (1949) *The Social Areas of Los Angeles: Analysis and Typology.* Westport, CO: Greenwood Press.

SHEVSKY, E. and BELL, W. (1955) *Social Area Analysis.* Stanford: Stanford University Press.

SILVER, J. and SHREVE, C. (1979) *Condominium Conversion Control.* Washington: U.S. Department of Housing and Urban Development.

SILVER, M. (1981) "You too can organize a government." *The Organizer* 9, no. 1 (Summer): 25-29, 32-37.

SIMON, D. A. (1979) *The Barter Book.* New York: E. P. Dutton.

SIMPSON, D., STEVEN, J., and KOHNEN, R. [ed.] (1979) *Neighborhood Government in Chicago's 44th Ward.* Champaign, IL: Stipes Publishing Co.

SKOGAN, W. G. and MAXFIELD, M. (1981) *Coping with Crime.* Beverly Hills, CA: Sage Publications.

SLAYTON, W. L. and DEWEY, R. (1953) "Urban Development and the Urbanite" in C. Woodbury, ed., *The Future of Cities and Urban Redevelopment.* Chicago: University of Chicago Press.

SMITH, C. J. (1978) "Self-help and social networks in the urban community." *Ekistics* 45 (March): 106-115.

SNEDEKER, B. B. and SNEDEKER, D. M. (1978) *CETA: Decentralization on Trial.* Salt Lake City: Olympus Publishing Co.

SOLOMON, S. (1979) *Neighborhood Transition without Displacement: A Citizens' Handbook.* Washington: National Urban Coalition.

SPIEGEL, H. B. C. [ed.] (1968) *Citizen Participation in Urban Development: Concepts and Issues* (vol. 1) *Cases and Programs* (Vol. 2). Washington: NTL Institute for Applied Behavioral Science.

——— (1977) "Evaluating program results," in H. W. Hallman (ed.) *The Organization and Operation of Neighborhood Councils.* New York: Praeger.

——— (1981) "Coalition of grassroots groups." *Citizen Participation* 2, no. 4 (March/April): 6.

SPIRO, H. R. (1969) "On beyond mental health centers: a planning model for psychiatric care." *AMA Archives of General Psychiatry* 21: 646-655.

SRI (1982) *Forming Urban Partnerships.* Menlo Park, CA: SRI.

STACK, C. (1974) *All Our Kin.* New York: Harper & Row.

STEGGERT, F. X. (1975) *Community Action Groups and City Government Perspective from Ten American Cities.* Cambridge, MA: Ballinger.

STEGMAN, M.A., ed. (1979) "Symposium on neighborhood revitalization" in *Journal of American Planning Association* 45, no. 4 (October): 458-556.

STEIN, B. (1974) *Size, Efficiency, and Community Enterprise.* Cambridge, MA: Center for Community Economic Development.

——— (1975) *Rebuilding Bedford-Stuyvesant: Community Economic Development in the Ghetto.* Cambridge, MA: Center for Community Economic Development.

STERNLIEB, G. and BURCHELL, R. W. (1973) *Residential Abandonment: The Tenement Landlord Revisited.* New Brunswick, NJ: Center for Urban Policy Research, Rutgers University.

STIPAK, B. (1980) "Local goverments' use of citizen surveys." *Public Administration Review* 40, no. 5 (September/October): 521-525.

STOECKLE, J. D. and CANDIB, L. M. (1974) "The neighborhood health center -- reform idea of yesterday and today" in R. Hollister et al. (eds.) *Neighborhood Health Centers.* Lexington, MA: Lexington Books.

STOKES, B. (1981) *Helping Ourselves: Local Solutions to Global Problems.* New York: Norton.

SULLIVAN, L. (1969) *Build Brother Build.* Philadelphia: Macrae Smith.

SUMMERFIELD, H. L. (1971) *The Neighborhood-Based Politics of Education.* Columbus, OH: Charles E. Merrill Publishing Co.

SUNDQUIST, J. L. and DAVIS, D. W. (1969) *Making Federalism Work.* Washington: Brookings Institution.

SUTTLES, G. D. (1968) *The Social Order of the Slum: Ethnicity and Territory in the Inner City.* Chicago: University of Chicago Press.

——— (1972) *The Social Construction of Communities.* Chicago: University of Chicago Press.

TAEBEL, D. (1978) "Minority representation on city council" in *Social Science Quarterly* 59 (June): 142-152.

TAUB, R. P. et al.; (1977) "Urban voluntary associations: locality based and externally induced" in *American Journal of Sociology* 83 (September): 425-424.

THERNSTROM, S. (1969) *Poverty, Planning and Politics in the New Boston: The Origins of ABCD.* New York: Basic Books.

THOMAS, W. I. and ZNANIECKI, F. (1920) *The Polish Peasant in Europe and America.* New York: Knopf.

THOMPSON, W. R. (1965) *A Preface to Urban Economics.* Baltimore: Johns Hopkins Press.

THRASHER, F. M. (1927) *The Gang.* Chicago: University of Chicago Press.

THURSZ, D. and VIGILANTE, L. [ed.] (1978) *Reaching People: The Structure of Neighborhood Services.* Beverly Hills, CA: Sage Publications.

TIMMS, D. E. D. (1971) *The Urban Mosaic: Towards a Theory of Residential Differentiation.* Cambridge, England: Cambridge University Press.

TOBIN, D. (1980) "Barter networks: working systems for self-help." *Exchange Networks* (July/August).

——— and FLETCHER, R. K. (1981) *Developing a Community-Based Barter Program.* Washington: VOLUNTEER: The National Center for Citizen Involvement.

TOENNIES, F. (1957) "Gemeinschaft and Gessellschaft," (1887), in E. P. Loomis, trans. and ed. *Fundamental Concepts of Sociology.* East Lansing, MI: Michigan State University Press.

TOMASIC, R. and FEELEY, M. M. [eds.] (1982) *Neighborhood Justice: Assessment of an Emerging Idea.* New York: Longman.

TOMEH, A. K. (1967) "Participation in a metropolitan community." *Sociological Quarterly* 8:85.

TRUMAN, D. (1951) *Governmental Processes: Political Interests and Public Opinion.* New York: Knopf.

Trust for Public Land (1979) *Citizen's Action Manual: A Guide to Recycling Vacant Property in Your Neighborhood.* Washington: Government Printing Office.

——— (1980) *Neighborhood Revitalization Manual.* Washington: National Park Service.

TURNER, J. B. [ed.] (1968) *Neighborhood Organizations for Community Action.* New York: National Association of Social Workers.

U.S. Bureau of the Census (1983) Unpublished data from 1982 census of Governments.

U.S. Department of Housing and Urban Development [HUD] (1967) *City Demonstration Agency Letter No. 3.* Washington: Author.

——— (1968)

Citizen Participation in Model Cities. Technical Bulletin No. 3. Washington: Author.

——— (1973) *Abandoned Housing Research: A Compendium.* Washington: Author.

———, Office of Policy Development and Research (1975) *Neighborhood Preservation: A Catalogue of Local Programs.* Washington: Author.

——— (1978a) *Citizen Participation in the Community Development Block Grant Program: A Guidebook.* Washington: Author.

———, Library Division (1978b) *Redlining: A Bibliography.* Washington: Author.

———, Library Division (1979a) *Neighborhood Conservation and Property Rehabilitation: A Bibliography.* Washington: Author.

———, Office of Neighborhood Development (1979b) *Neighborhoods: A Self-Help Sampler.* Washington: Government Printing Office.

———, The Secretary of (1979c) *Displacement Report.* Washington: Author.

———, Office of Policy Development and Research (1981) *Residential Displacement: An Update.* Washington: Author.

———, Office of Public Affairs (n.d.) *The Urban Fair: How Cities Celebrate Themselves.* Washington: Author.

U.S. Office of Consumer Affairs, Consumer Information Division (1979) *People Power: What Communities Are Doing to Counter Inflation.* Washington: Author.

U.S. Office of Economic Opportunity [OEO] (1966) *The Neighborhood Center.* Washington: Author.

Urban Land Institute (1964) *The Homes Association Handbook.* Washington: Author.

Urban Systems Research and Engineering, Inc. (1977) *The Urban Homesteading Catalogue* 3 vol. Washington: U.S. Department of Housing and Urban Development.

———, (1979) *The Role of Community Based Organizations under CETA.* Washington: U.S. Department of Labor.

——— (1980) *Neighborhood Planning Primer.* Washington: U.S. Department of Housing and Urban Development.

——— (1981) *The Role of Nonprofit Organizations in CETA Public Service Employment.* Washington: U.S. Department of Labor.

VERBA, S. and NIE, N. H. (1972) *Participation in America: Political Democracy and Social Equality.* New York: Harper & Row.

VIGILANTE, J. L. (1976) "Back to the old neighborhood." *Social Service Review* 50 (June): 194-208.

WALDHORN, S. A. and GOLLUB, J. O. (1980) *Using Nonservice Approaches to Address Neighborhood Problems: A Guide for Local Officials.* Menlo Park, CA: SRI International.

WALLACE, A. F. C. (1952) *Housing and Social Structure.* Philadelphia: Philadelphia Housing Authority.

WANDERSMAN, A. (1981) "A framework of participation in community organizations." *Journal of Applied Behavioral Science* 17:27-58.

———, JAKUBS, J., and GIAMARTINO, G. (1981) "Participation in block Organizations." *Journal of Community Action* 1, no. 1 (September/October): 40-47.

WARNER, W. L., MEEKER, M., and EELS, K. (1949) *Social Class in America: A Manual of Procedure for the Measurement of Social Status.* Chicago: Science Research Associates.

WARREN, D. I. (1981) *Helping Networks: How People Cope with Problems in Urban Communities.* Notre Dame, IN: University of Notre Dame Press.

WARREN, R. B. and WARREN, D. I. (1977) *The Neighborhood Organizer's Handbook.* Notre Dame, IN: University of Notre Dame Press.

WARREN, R. L. (1978) *The Community in America.* Chicago: Rand McNally.

——— [ed.] (1977) *New Perspectives on the American Community: A Book of Readings.* Chicago: Rand McNally.

Washington Consulting Group (1974) *Uplift: What People Themselves Can Do.* Salt Lake City: Olympus Publishing Co.

WASHNIS, G. J. (1972) *Municipal Decentralization and Neighborhood Resources: Case Studies of Twelve Cities.* New York: Praeger.

——— (1974) *Community Development Strategies: Case Studies of Major Model Cities.* New York: Praeger.

——— (1976) *Citizen Involvement in Crime Prevention.* Lexington, MA: Lexington Books.

WATMAN, W. S. (1980) *A Guide to the Language of Neighborhoods.* Washington: National Center for Urban Ethnic Affairs.

WATSON, F. (1922) *The Charity Organization Movement in the U.S.: A Study in Philanthropy.* New York: Macmillian.

WEATHERFORD, M. S. (1982) "Interpersonal networks and political behavior." *American Journal of Political Science* 26, no. 1 (February): 117-143.

WEBB, K. and HATRY, H. P. (1972) *Obtaining Citizen Feedback: The Application of Citizen Surveys to Local Government.* Washington: Urban Institute.

WEBER, N. and LUND, L. (1981) *Corporations in the Community: How Six Major Firms Conduct Community Programs.* New York: The Conference Board.

WEGENER, E. (1979) *Community Crime Prevention.* Washington: Civic Action Institute.

WEILER, C. (1978) *Reinvestment Displacement: The Government Role.* Washington: National Association of neighborhoods.

WEIS, J. F. et al. (1980) *Dividing the Pie: Resource Allocation to Urban Neighborhoods.* Boston: Neighborhood Development Agency.

WEISS, C. H. (1972) *Evaluation Research: Methods of Assessing Program Effectiveness.* Englewood Cliffs, NJ: Prentice-Hall.

WEISSMAN, H. H. (1970) *Community Councils and Community Control: The Workings of Democratic Mythology.* Pittsburgh: University of Pittsburgh Press.

WEISSMAN, S. R. (1978) "Limits of citizen participation: lessons from San Francisco's model cities program." *Western Political Quarterly* 31, no. 1 (March): 32-47.

WELLMAN, B. and LEIGHTON, B. (1979) "Networks, neighborhoods, and communities: approaches to the study of the community question." *Urban Affairs Quarterly;* 14, no. 3 (March): 363-390.

WERTH, J. T. and BRYANT, D. (1979) *A Guide to Neighborhood Planning.* Chicago: American Planning Association.

WHITAKER, G. P. (1980) "Coproduction: citizen participation in service delivery." *Public Administration Review* 40, no. 3 (May/June): 240-246.

WHITE, C. R. and ENDER, S. M. (1981) "Participation in neighborhood organizations." *Journal of Community Action* 1, no. 1 (September/October): 48-52.

WHYTE, W. F. (1943a) *Streetcorner Society.* Chicago: University of Chicago Press.

——— (1943b) "Social organization in the slums." *American Sociological Review* 8: 34-39.

WHYTE, W. H., Jr. (1956) *The Organization Man.* New York: Simon & Schuster.

WILENSKY, H. L. and LEBEAUX, C. H. (1958) *Industrial Society and Social Welfare.* New York: Russell Sage Foundation.

WILLIAMS, J. A., Jr., BABCHUK, N., and JOHNSON, D. R. (1973) "Voluntary association and minority status: a comparative analysis of Anglos, blacks, and Mexican-American." *American Sociological Review* 38 (October): 637-646.

WILSON, J. Q. (1963) "Planning and politics: citizen participation in urban renewal." *Journal of the American Institute of Planners* 29, no. 4 (November).

——— (1973) *Political Organization.* New York: Basic Books.

WILSON, R. A. (1971) "Anomie in the ghetto: a study of neighborhood type, race, and anomie." *American Journal of Sociology* 77, no. 1 (July): 66-68

WILSON, R. L. (1962) "Liveability of the city: attitudes and urban development," in F. S. Chapin, Jr., and S. F. Weiss, (eds.) *Urban Growth Dynamics in a Regional Cluster of Cities.* New York: John Wiley.

WIRTH, L. (1928) *The Ghetto.* Chicago: University of Chicago Press.

——— (1938) "Urbanism as a way of life." *American Journal of Sociology* 44 (July): 1-24.

WOLFINGER, R. E. and ROSENSTONE, S. J. (1980) *Who Votes?* New Haven, CO: Yale University Press.

WOOD, R. C. (1958) *Suburbia — Its People and Their Politics.* Boston: Houghton Mifflin.

WOODSON, R. L. (1981) *A Summons to Life: Mediating Structures and the Prevention of Youth Crime.* Cambridge, MA: Ballinger.

——— (1982a) "The importance of neighborhood organizations in meeting human needs," in J. A. Meyer (ed.) *Meeting Human Needs.* Washington: American Enterprise Institute.

——— (1982b) "Youth crime prevention: an alternative approach," in J. A. Meyer (ed.) *Meeting Human Needs.* Washington: American Enterprise Institute.

Woodstock Institute (1982) *Evaluation of the Illinois Neighborhood Development Corporation.* Washington: U.S. Department of Housing and Urban Development.

WOODY, B., WALTERS, R. W., and BROWN, D. R. (1980) "Neighborhoods as a power factor." *Society* 17, no. 4 (May/June): 49-55.

YANCEY, W. (1971) "Architecture, interaction and social control: the case of a large-scale housing project." *Environment and Behavior* 3: 3-21.

——— and ERICKSON, E. P. (1979) "The antecedents of community; the economic and institutional structure of urban neighborhoods." *American Sociological Review* 33 (April): 253-266.

YATES, D. (1973) *Neighborhood Democracy: The Politics and Impacts of Decentralization.* Lexington, MA: Lexington Books.

YIN, R. K. (1982) *Conserving America's Neighborhoods.* New York: Plenum Press.

YIN, R. K. et al. (1973) *Citizen Organizations: Increasing Client Control over Services.* Santa Monica, CA: Rand.

YIN, R. K. and YATES, D. (1974) *Street-level Governments: Assessing Decentralization and Urban Services.* Santa Monica, CA: Rand.

YIN, R. K., VOGEL, M. E., CHAIKEN, J. M., and BOTH, D. R. (1976) *Patrolling the Neighborhood Beat: Residents and Residential Security.* Santa Monica, CA: Rand.

ZERCHYKOV, R. and DAVIES, D. (1980) *Leading the Way: State Mandates for School Advisory Councils in California, Florida, and South Carolina.* Boston: Institute for Responsive Education.

——— (1982) *Citizen Participation in Education: Annotated Bibliography.* Boston: Institute for Responsive Education.

ZIMMERMAN, J. F. (1972) *The Federated City: Community Control in Large Cities.* New York: St. Martin's.

ZITO, J. M. (1974) "Anonymity and neighboring in an urban high-rise complex." *Urban Life and Culture* 3: 243-263.

ZORBAUGH, H. W. (1929) *The Gold Coast and the Slum.* Chicago: University of Chicago Press.

ZURCHER, L. A. (1970) *Poverty Warriors.* Austin: University of Texas Press.

INDEX

ABOUT THE AUTHOR

Howard W. Hallman is President of the Civic Action Institute, a private nonprofit organization he founded in 1969 (originally known as the Center for Governmental Studies). He has worked for citizen organizations and local government in Philadelphia (1952-1958) and New Haven (1959). He functioned as a free-lance consultant from a base in Washington, D.C., from 1965 to 1969, interrupted in 1967 for a year of service directing a Senate subcommittee study of the War on Poverty. Hallman received bachelor and master degrees in political science from the University of Kansas, his native state. His books include *Community Programs on Employment and Manpower* (1966), *Neighborhood Control of Public Programs* (1970), *Neighborhood Government in a Metropolitan Setting* (1974), *Emergency Employment: A Study in Federalism* (1977), *The Organization and Operation of Neighborhood Councils: A Practical Guide* (1977), *Small and Large Together: Governing the Metropolis* (1977), and *Community-Based Employment Programs* (1980).